SLANTED

Also by Sharyl Attkisson

Stonewalled: My Fight for Truth Against the Forces of Obstruction, Intimidation, and Harassment in Obama's Washington

The Smear: How Shady Political Operatives and Fake News Control What You See, What You Think, and How You Vote

SLANTED

HOW THE NEWS MEDIA TAUGHT US
TO LOVE CENSORSHIP
AND HATE JOURNALISM

SHARYL
ATTKISSON

HARPER
An Imprint of HarperCollinsPublishers

HarperCollins books may be purchased for educational, business, or sales promotional use. For information, please email the Special Markets Department at SPsales@harpercollins.com.

FIRST EDITION

Library of Congress Cataloging-in-Publication Data
Names: Attkisson, Sharyl, date.
Title: Slanted: how the news media taught us to love censorship and hate journalism / Sharyl Attkisson.
Description: First edition. | New York City: Harper, 2020. | Includes index.
Identifiers: LCCN 2020030750 (print) | LCCN 2020030751 (ebook) | ISBN 9780062974693 (hardcover) | ISBN 9780062974709 (ebook)
Subjects: LCSH: Journalism—Political aspects—United States—History—21st century. | Press and politics—United States—History—21st century. | Journalism—Objectivity—United States. | Journalistic ethics—United States.
Classification: LCC PN4888.P6 A85 2020 (print) | LCC PN4888.P6 (ebook) | DDC 071/.3—dc23
LC record available at https://lccn.loc.gov/2020030750
LC ebook record available at https://lccn.loc.gov/2020030751

20 21 22 23 24   LSC   10 9 8 7 6 5 4 3 2

With love and gratitude to my friends, family,
attorneys, and my other partners in truth.
Some of the proceeds from Slanted *are being donated to the*
University of Florida College of Journalism and Communications,
and other good journalism and anti-censorship causes.
The content of this book is based on my own opinions, experiences,
and observations. Some quotes contained within are based on
my best recollections of the events and, in each instance,
accurately reflect the spirit and my sense of the conversations.

In a time of deceit, telling the truth is a revolutionary act.

—UNKNOWN

Logic is an enemy and truth is a menace.

—"THE OBSOLETE MAN"

Contents

SLANTED

Introduction

In George Orwell's dystopian novel *1984*, the hapless protagonist, Winston Smith, is a government records editor at the Ministry of Truth—which is a job that's all about lies.

Poor Winston's assignment is to painstakingly rewrite history in real time. Revise old newspaper stories to make them line up with the ruling political party's current version of the truth. It's a job that never ends. History must constantly be altered because one lie inevitably necessitates another. And the needs of Big Brother—the dictator in this totalitarian society—require that a position declared one day be erased and forgotten the next.

To accomplish its goal, Big Brother mandates the destruction of all paper records. The citizenry must deposit any surviving documents into "memory holes," never to be referenced again. There isn't any real "news"—only that which the powerful decide people should hear and believe: the censored, curated, and sanitized.

Today, we're in an Orwellian environment that has taken this frightening scenario a step further. Big Brother constantly revised "facts" to fit the government's ever-changing story. The modern media have also discovered how to carefully filter information on the front end to make sure that only the "correct" view is presented in the first place. That way, the story never has to change.

Right now, as you read these words, versions of history and current events are being written and revised in real time according to what powerful interests wish them to say. Our "memory hole" is found in growing efforts to "curate" or censor information on the news, ban certain facts, declare selected viewpoints illegitimate, cleanse social media of particular accounts, and judge people and events of the distant past using today's evolving and controversial standards.

Even those who know better are left, like Winston Smith, to guess and wonder *how many others like them* are out there—how many of the unindoctrinated who don't buy the spin? There's certainly no way to find out by clicking on different Internet articles or flipping among cable news channels.

This giant purge of knowledge and facts wouldn't be possible without the news media. We in the media have, to a frightening degree, gotten on board with the efforts to convince the public that they do not need or deserve access to all information, only that which powerful interests see fit for them to have.

Reporters are so aware of this that they have a name for it: The Narrative. The phrase is used to describe what we caught *others* doing to try to shape the news. Now we're doing it ourselves.

The Narrative refers to a story line that influential people want told in order to define and narrow your views. The goal of The Narrative is to embed chosen ideas so deeply within society that they are no longer questioned—scratch that—so that questions are not permitted.

Slanted tells the story of what happens when reporters convince news consumers that the reporters' own opinions are more valuable than facts. With an information universe at our fingertips on the news and Internet and with propagandists working overtime to shape it, many people ask what they can believe. Journalists are more than happy to tell them. Unfortunately, the journalists are too often driven by propaganda, as well.

The goal of this book is to help you expose and defeat narratives even when they are cleverly executed by the most powerful sources using the most sophisticated methods. It will also reveal how the business of narratives is inextricably linked to the death of the news as we once knew it.

I will anatomize a series of narratives that have dominated even in the face of contradictory facts. Anyone accused of sexual harassment must be guilty if there are enough accusers, no matter how flimsy the claims may be. Donald Trump is too cowardly to visit the troops in a war zone. When "mass shootings" occur in certain cities, they must be called something else. Russia changed the outcome of the 2016 election. All new polling spells doom for Republicans. And many more.

The point is that The Narrative is guiding what facts you get to learn about. Facts that serve The Narrative are deemed to be "news." Facts that don't are not news. Or are to be obliterated.

Defining The Narrative

To begin with, a narrative almost always presents multisided issues in a distinctly one-sided fashion. Any notion of logic is suspended. The standards and judgments applied to the target being smeared by a narrative are never applied to those advancing the narrative or their allies. For example, someone pushing a narrative might accuse his target of lying or being hateful or racist. At the same time, the one doing the accusing may be lying or acting in a hateful or racist way—but no attention is given to the hypocrisy. People simply pretend to not notice. You'll see a lot of real-life examples in this book.

You might think that a defining characteristic of a narrative is that it is false. But that's often not the case. Here are three ways in which truthful information can also qualify as narratives.

First, when truthful information is deliberately presented in a biased fashion in order to confuse, drown out, or overwhelm other facts and to advance a particular goal. For example, it may be true that a mass killer used a gun. But news reports about the crime serve a narrative if they are overwhelmingly shaded to the exclusion of counterpoints in order to make an argument for gun control.

Second, truthful information can qualify as a narrative when it is amplified beyond its independent news value in order to promote a broader story line. For example, it may be true that former first lady Hillary Clinton stumbled when descending a set of stairs. But news reports on such an incident serve a narrative if they become front-page headlines and a trending topic on social media to imply, absent other hard evidence, that Clinton's stumble proves she's seriously ill.

And third, the truth can become a narrative when it is couched in terms that present an issue as a closed case never to be reopened or implies that contrary facts and views are illegitimate. For example, there may be a good reason to discuss the frequency of tornadoes or rising floodwaters theoretically in terms of global warming. But the discussion becomes a narrative if news analysts link every weather phenomenon to man-made climate change, as if it is a fact, with little consideration given to scientific counterpoints.

Once a narrative is successfully established, a great deal of effort must be put into cultivating it. Contrary views, facts, and science must be shoved down the memory hole—disappeared—as though they had never existed.

Accomplishing this propaganda feat in the information age re-quires a great deal of coordination. That includes campaigns to con-vince the public at large to embrace the once unthinkable notion that their news should be curated by third parties. It includes well-funded "media literacy" efforts to brainwash—er, teach—us and our children whom to believe and whom to tune out. It includes infiltrat-ing our universities and public schools. It includes proposing laws that promote censorship and turn free speech on its head, creating policies that result in narrowing the universe of available informa-tion, and plain old bullying of those who don't obediently dance in step behind the appointed Pied Piper.

The news is being used to accomplish all of these things.

When "the news" is utilized to further narratives, it requires us to deviate seriously from fact-based reporting. The Narrative may require that information be presented in a slanted fashion or that facts be taken out of context. And, of course, it may involve reporting entirely false material. Unfortunately, that's become quite the trend. And that's perhaps the biggest modern victory of The Narrative.

There is an important component of The Narrative when it advances political interests: it is always presented as nonpolitical. Any version of events that counters The Narrative is called partisan spin. An article filled with anonymous sources about a government investigation is a potential Pulitzer Prize winner if it supports The Narrative. If it does not, it's portrayed as a partisan hit job.

It is important to recognize that the people behind a narrative do not always have cynical or evil motives. They may even be acting according to what they believe to be a higher purpose. In such cases, these people share an important belief: that they are smarter than you are. They do not trust you to process information and draw your own conclusions because you might draw the wrong ones. You must not be left to your own devices. So, much like Big Brother, they dictate which views are to be considered legitimate and which are off-limits. They tell you what to think. They become the ultimate arbiters of truth even when it's a matter of debate or opinion. It's all for your own good.

Settled science. Not open to debate. Everyone agrees.

The Psychology of The Narrative

A hefty deployment of "doublethink"—described in *1984* as a tactic to psychologically manipulate the citizenry—is helpful in understanding the psychology of The Narrative. Orwell defined doublethink as "To know and not to know, to be conscious of complete truthfulness while telling carefully constructed lies, to hold simultaneously two opinions which canceled out, knowing them to be contradictory and believing in both of them . . . to forget whatever it was necessary to forget, then to draw it back into memory again at the moment when it was needed, and then promptly to forget it again, and above all, to apply the same process to the process itself."

News reporters and pundits must accept doublethink in order to service The Narrative with a guilt-free conscience. Then they must condition news consumers to use doublethink to reject intellect and reason. All must become unquestioning of The Narrative; accepting of the antithetical; skeptical of the "wrong" people. The public must be conditioned to attack those who try to shed new light on an issue or have a dialogue about it, and must pledge zealous support to the ones who are actually fooling them.

The existence of The Narrative explains the otherwise inexplicable. The Narrative is why, when there are thousands of news topics that could be dissected, we see the same relative handful of stories repeated on the national news day in and day out.

The Narrative is also why we see the same faces on the national news over and over again, no matter how unreliable or inaccurate they've proven to be. It explains why reporters continue to consult the usual suspects, even after they have provided false information. It answers the question of how news organizations can rely on analyses from former government intelligence officials commenting on issues for which they themselves are under investigation—even after they have repeatedly been proven wrong.

There's no better example than the political operative Donna Brazile. You may recall that in 2016, the CNN contributor and Democratic Party official was caught passing along inside news information to the Hillary Clinton campaign, then denied having done so. She was fired from CNN after her deeds became public. If the true purpose of political pundits as media contributors were to provide honest analysis and meaningful opinions, Brazile would be banished from TV studios for life (except, perhaps, to address the controversies surrounding herself).

The Narrative explains why Brazile didn't suffer that fate—quite the opposite. Instead, the media offered a heartfelt embrace after her unfortunate episode. She was invited to appear on other news networks to opine on and provide the Democrats' views, with reporters and pundits sitting next to her politely pretending—quite convincingly—that she had not been discredited. When furthering a narrative is the goal, truth, accuracy, and reliability take a back seat. Only in this environment does Donna Brazile's inexplicable trajectory make sense.

The Narrative is not solely the invention of political figures; corporate interests are masters at inventing narratives that exploit the lucrative synergy between business and news. Narratives that benefit corporations are adopted by a conflicted media thirsty for sponsorships and ad dollars. The news can become little more than a distribution tool for the corporate narrative.

What happens to news reporters who are off narrative? They suffer the full wrath of the Narrative establishment. They may be bullied, attacked, shouted down, investigated, sued, researched, controversialized, and slandered with every available propaganda tool.

A popular narrative today is that Donald Trump is responsible for killing the news as we once knew it. After all, he threatened to open up libel laws to make it easier to sue the press. He led chants against the "fake news" at his campaign rallies. His staff wrestled a microphone away from and temporarily banned a CNN reporter from White House briefings. Trump labeled fake news the "enemy of the American People." *How can we in the news be expected to remain fair and neutral? Why should we maintain a professional distance?* After all, Trump made it personal.

But what if the off-narrative version tells a different story?

Through his unconventional ways that defied predictions and operated outside the controlling narratives, Trump exposed bias, flaws, and weaknesses in the news media, causing its members to lose their collective mind and shed all pretense of objectivity. The media at large became committed to a political agenda to undermine and ultimately remove Trump from office. Which only served to prove his point about their bias.

Within these pages, I'll make liberal use of the "Substitution Game"

I devised to demonstrate the media's disparate treatment of topics and people in order to fit a particular narrative.

You'll hear candid opinions and analyses—some of them startling—from dozens of top news executives, reporters, and producers currently or formerly employed by CNN, CBS, NBC, ABC, MSNBC, Bloomberg News, and the *New York Times*. They work (or worked in the past) on programs such as *PrimeTime Live*, *Nightline*, and *60 Minutes*. They're people who have rubbed shoulders with such notables as Walter Cronkite, Ted Koppel, Katie Couric, and Diane Sawyer. Many were eager to share their opinions on news narratives and "the death of the news." Many did not wish to be quoted by name so that they could speak freely and not be ostracized for critiquing their own industry and colleagues. Of those who told me where they personally stand in terms of politics, none referred to themselves as conservative. Most said they consider themselves liberal, progressive, or very liberal. One told me he considers himself "pretty much down the middle." Most described their political views in ways such as "not extreme" or "not militantly one-sided."

This autopsy will prove that the death of the news as we once knew it isn't an act of murder but suicide. And The Narrative was the weapon.

In *1984*, the government's Ministry of Peace conducts war. The Ministry of Love deploys cruel punishment. The Ministry of Truth falsifies historical records.

In 2020, we have our own versions:

FACT-CHECKERS CODIFY SLANTED OPINION.

MYTH BUSTERS DISPEL TRUTH.

ONLINE KNOWLEDGE IS SHAPED BY AGENDA EDITORS.

FREE SPEECH IS CONTROLLED BY CENSORSHIP.

THE NEWS—ISN'T THE NEWS.

AND YOU AREN'T THE CONSUMER; YOU'RE THE PRODUCT.

Eventually, as told in *1984*, the masses lose the ability to form independent thoughts. The Party can convince them that anything is true.

This book will serve as an enduring resource for independent thinkers. It will expose—in painstaking detail—the complex web of narratives we encounter every day. And we will find sparks of hope that provide reason for optimism—one of which is the fact that you're reading these words.

CHAPTER 1

CBS Tales: "Death by a Thousand Cuts"

Come with me on the early days of my journey. Hear how I came to realize that there are two harmful types of slant in news reporting: bias that is intentional, and that which is unwitting.

Intentional bias, as audacious as it is, is almost easier to address. It is worn on one's sleeve. It is proud and undeniable. Reporters usually know when they are committing it but convince themselves the bias is justified or the victim of the bias deserves it.

But unintentional bias . . . well, that's a sneaky little man. And much of the problem with the news today can be blamed on him. Corralling unintentional bias is like trying to cling onto smoke or sinking your teeth into a heaping bite of water. Because the mere idea that bias exists escapes those who are displaying it. They fall victim to their own bias even as they believe they cannot be guilty of it. Sometimes we recognize unintentional bias in our colleagues or bosses. But calling them on it or appealing to logic fails to convince them to reconsider their worldview. Instead, they may look at you askance as if to say, *What's wrong with* you?

Obviously, bias can be political. But you may be surprised to learn how much of it has little to do with politics.

The push-me-pull-you over bias deserves constant attention in America's newsrooms. Yet it receives little attention in most. And this conflict impacts how the newsrooms work, the news they cover, the trust they earn among readers and viewers. Even when there are people in the newsroom who reach out and try to grab onto the smoke, that sneaky little man manages to triumph.

The seeds that enable this dynamic were planted long ago. They have sprouted and grown for many years.

In 1996, while I was working as a correspondent for *CBS Weekend News* in Washington, DC, an assignment came down from New York: *Do a story on why Steve Forbes's flat tax won't work.* Forbes was a Republican candidate running for president. The assignment, as worded, assumed a prejudged conclusion. The Narrative, simply put, was that Forbes's flat tax would benefit the rich and hurt the poor.

In fact, nobody knows for sure what the impact of a flat tax would be. Economists differ, and certainly reporters cannot claim to know for sure. The assignment should have asked me to explore both sides of whether a flat tax would work. So that was how I set out to execute the story. Even back then, it struck me that here I was working at a prestigious national news network, but some of my experienced colleagues seemed to have no recognition that they were operating with a blatantly slanted mentality. Sometimes we became so focused on how we think a story was supposed to come out that we missed the real news.

Many years after that incident, not long before I left CBS in 2014, a young colleague popped into my office at 2000 M Street, NW, in Washington, DC. She told me she had been assigned to do a story about the importance of food pantries for poor people who must rely on them. Over the course of the next week, she kept me updated on her frustrating quest to find the right family to profile. Every food pantry family she had connected with had proven to be relatively well off financially. They didn't fit the bill of being desperately poor. In one case, she told me, a food pantry recipient had invited her to shoot video of him and his family at home for the news story. Once there, my colleague asked to look in the refrigerator with the camera, naturally expecting the cupboard to be bare. To her surprise, it was full of food—not just people-food but food for household pets, too. That would not make a sympathetic focus for her assignment.

By the time my colleague made her third visit to my office recounting her inability to find a very poor family relying on food pantries, I gently suggested that maybe the truth of the story was different from the one she had been assigned. Perhaps there was a story to be told about the type of people who were, in fact, visiting food pantries. People who were not totally destitute were turning to food pantries for help. What was *their* story? Why not report what she was actually learning from her experience in the field? Why force a narrative that in practice did not seem to exist? She looked at me as though I had grown a second head.

What she did not know was that the advice came from my own experience from when I was a younger journalist.

In the late 1990s, during the second term of the Clinton presidency, Labor Secretary Robert Reich was advocating for an increase in the national minimum wage. As part of that, he was quoted in a newspaper article lamenting that tens of thousands of families were trying to raise their children on minimum-wage incomes. The New York CBS office assigned me to find and profile one of those struggling families for a story to appear on *CBS Weekend News*. I did not look at it that way at the time, but in retrospect, I was being assigned to fulfill the preconceived narrative that there were large numbers of hardworking couples trying to raise their families on inadequate income due to greedy employers who needed to be forced to do the right thing; therefore, the minimum wage should be raised.

I set about trying to fill the order. I thought it would be easy. *So many people are raising families on minimum wage!* I just needed to find one. There are advocacy groups for just about everything in Washington, DC, and they are more than happy to make the job of a reporter under deadline quicker and easier. I turned to them to help find the example we were looking for.

I was surprised when some time passed and none of the advocacy groups came up with a family for me to interview. When I followed up with them, the advocates confessed that they had not been able to locate any families where two parents were raising kids while earning a minimum wage.

"Okay," I said, *"how about a couple with kids where one parent is earning minimum wage?"* That, too, turned out to be a no-go. They could not find such a family. I modified the request again: *How about a single parent raising kids on minimum wage?* They scoured their contacts and came up empty again. *How about a couple with no kids trying to scrape by on minimum wage?* Nada.

I was so bent on trying to fulfill the assignment as given that I missed the forest for the trees. Perhaps the real story was that even the most motivated and well-connected advocacy groups promoting a higher minimum wage couldn't identify a single one of the "tens of thousands" of American families who were supposedly raising their children on minimum wage.

My last request to those advocates felt lame. I asked if they could just find me a single person with no kids who was living on minimum wage. While I waited, I decided to take matters into my own hands. I embarked upon my own search for minimum-wage families. *How hard can it be?* I thought. *After all, there are tens of thousands! The government says so!*

I contacted restaurants, pizza delivery chains, dry cleaners—anyplace I could think of that I assumed would be paying minimum wage. I quickly learned that most states, as well as many cities and counties, have a higher minimum wage than what federal law requires. I also learned that, yes, some minimum-wage jobs exist in the summertime when college students are home looking for seasonal work. But it wasn't summer at the time. And college kids would not have been good examples for the story because they were not supporting themselves or raising children on a minimum wage.

I still didn't give up. There must be someplace that could deliver me a minimum-wage family. Maybe McDonald's! Surely, the cheapest fast-food restaurant I knew of must pay minimum wage. So I walked from my office up M Street in the Northwest section of Washington, DC, to a nearby McDonald's. I'd become friendly with the manager there during my frequent visits for a midafternoon Coke. Surely he would be able to connect me with a minimum-wage worker.

As I explained my assignment, the manager started shaking his head before I even finished. First, he explained, Washington, DC, is one of the places that has a higher minimum wage than federal law requires. (In 2020, for example, the Washington, DC, minimum wage was $15.00, more than double the federal minimum wage of $7.25 an hour.)

"I'll be honest with you," the manager told me. "Even if I had someone starting at minimum wage, they don't stay there. If they just show up for work every day, they get a twenty-five-cent raise every three months. Nobody here is living on a minimum-wage salary."

Maybe that was the real story, all of the facts I learned that caught me by surprise: how many locales pay above the federal minimum wage; how businesses that I had assumed paid the lowest hourly rate actually pay more; how difficult it was to find anyone raising children on minimum wage. Now, *that* was interesting! But it did not occur to me to suggest changing the story assignment to reflect what I'd actually found in the field.

Meantime, the last advocacy group still trying to help me finally got back to me. "The only thing we can offer is an elderly, retired man in Maryland who, by choice to keep busy, works cleaning public parks for minimum wage," its representative told me. It was a far cry from representing tens of thousands of families. But, you guessed it, that senior citizen became the centerpiece of my story. I unwittingly bent myself into a pretzel to deliver the predetermined narrative.

I would conduct many similar pursuits throughout my network career before I started to have an awakening: too often, we in the news try to serve up The Narrative instead of the facts.

In 2004, a senior producer at CBS News assigned several of us to choose and explore a campaign issue through one "character."

A number of us thought the assignment was fraught with peril. Here's why. Picking a single person to explore a political controversy invariably creates empathy for the side of the chosen character. It is unlikely to produce an evenhanded news story. For example, let's say the chosen issue is abortion. Invariably, the character selected is a woman who needed to terminate a pregnancy to save her life or a rape victim who did not want to carry a pregnancy to term. Profiling such a woman would naturally create a slanted story generating sympathy for her and the pro-choice side with no fair counterpoint. Adding to my worries about the appearance of bias was the fact that we were already facing down quite a bit of public criticism at CBS about our supposed liberal bias.

I called the senior producer who assigned the stories. I explained that for our own good at CBS, to protect our reputation for fairness, we should produce stories that explored *both* sides of any given controversy and interview "characters" or people representing each side. But the senior producer dismissed my concerns. Her reasoning was that various producers and correspondents would pick different sides of issues; some would tilt liberal, others conservative, and it would all somehow even out in the wash.

I knew that would not be the case. For example, I was confident that no reporter or producer would choose to profile a pro-life woman who was happy that she had not aborted a child. The liberal viewpoint was going to be chosen in nearly if not every case. In any event, I moved forward with my assignment. I was determined to try to do my part to mitigate what the public would surely perceive as a liberal tilt among this high-profile featured group of election stories. I chose the topic of religion in schools.

You have to understand that the typical way religion in schools is covered on the news is by highlighting a case where someone is fighting to remove Christian references or the mention of God. One example would be the father who fought a court battle so that school students did not have to say the words "under God" in the Pledge of Allegiance.

But I didn't want to go the typical route. First, that story had already been well told by many others. I like to try to find different or underserved examples. Second, it would be unexpected, interesting, and off narrative to cover the story in a different way. Third, I had seen recent poll numbers indicating that as many as nine in ten

Americans agreed that the words "under God" belong in the pledge. Ninety percent of Americans also said that they believe in God. Although I am not personally religious, I recognize that religious Americans are a big part of the news audience that we do not typically serve well. It made sense from a viewership standpoint to tackle at least one story in a way that spoke to them.

The Washington, DC, producer I worked with found a good idea for who could become the central "character" in our story. Instead of an individual, it would be a small group of teens who held a Bible study group at a Maryland high school. They met on campus while following rules to maintain the constitutional separation between church and state. For example, although the Bible sessions were held on campus, they were convened after school hours, they were student organized, and no one was pressured to attend.

I then covered an opposing view by interviewing Annie Laurie Gaylor, the leader of a national group of atheists and agnostics. And I summarized the case of the California dad who fought to remove "under God" from the Pledge of Allegiance. Last, I interviewed someone who had a fairly neutral take: Charles Haynes of the First Amendment Center.

After I finished writing the story, I turned it in to the New York senior producer who'd made the assignment. She gave it her stamp of approval. But the next day, I got an unusual follow-up call. She'd changed her mind. She did not like the idea of basing the story on the high school Bible study kids.

"Can't you find someone with a more extreme position" to tackle the issue? she asked. "Try to find someone who wants to insert religion into the public school curriculum."

I found it an odd request. I told her that there was no serious push by anybody to "insert religion into public schools," so that would not be a good idea for our story.

"What about Jerry Falwell?" she pressed. "Can you interview Jerry Falwell?"

Falwell, a televangelist and Christian conservative activist, was not part of a push to insert religion into public schools. It suddenly became clear to me that this senior producer was looking for ways to create a flash point rather than cover a genuine issue. Her desired narrative, I suspected, was that people who embrace religion are unreasonable extremists—not like the high school students I had interviewed, who to our viewers would seem reasonable and likable.

"I'm not going to interview Jerry Falwell," I replied. "The story is good as it is."

With that, my report promptly fell off the *CBS Evening News* schedule. But that wasn't the end of it.

At the time, I was frequently filling in as anchor of *CBS Weekend News* in New York. Since the weekday evening news did not want the religion-in-schools story, I scheduled it to air on one of my upcoming weekend news broadcasts. Then, a few days before the airdate, I was in the offices at the CBS Washington, DC, bureau, and I walked past the desk of a producer on speakerphone with the weekend news executive producer and staff in New York. They were meeting to discuss the stories scheduled for the upcoming weekend.

"The anchor," said the executive producer, referring to me, "has a pro-Bible story she wants to air on Sunday." There was a sniff of clear distaste in her voice.

I was pretty surprised. My story was certainly not "pro-Bible." As I have described, there was nothing in the story that advocated for anything one way or another. I had fairly presented various views. Had I chosen to profile the boy whose father fought to remove "under God" from the Pledge of Allegiance, I am certain that the same executive producer would not have called it an "anti-Bible" piece. They just weren't used to a correspondent veering from the expected narrative. I could tell that it had ruffled feathers. Such reporters are to be viewed with skepticism—even suspicion.

An hour before anchoring the newscast in New York, I asked a senior producer on the broadcast why my story had inaccurately been described to the staff as "pro-Bible" when it was no such thing.

"Well, because I think religion is at the root of all evil and the cause of all wars!" he sputtered, seeming surprised that I had asked.

"What does your personal opinion, or mine for that matter, have to do with anything?" I countered.

Another moment of clarity for me came later that year. *CBS Evening News* assigned me to try to dig up dirt on President George W. Bush, who was running against Democrat John Kerry in the 2004 race. A lot of rumors were circulating about Bush's supposed (but unproven) cocaine use long before he was president.

I'm a pretty good digger, but I doubted that in a matter of a few days, I would be able to uncover evidence of Bush scandals that had eluded devoted Bush-hating reporters who'd spent a big chunk of their career looking for it.

Still, I set out to see what I could learn. I got up to speed on the rumors and allegations. There wasn't much to bite into. I then developed a working theory, which is sometimes a good way to find a starting point when investigating something vague. I figured that if Bush had truly had a serious drug and/or alcohol problem in his youth, his family might have sent him to a rehab center somewhere outside their home state of Texas. I figured it would have been a center where wealthy families could count on discretion. And I theorized that such a center would have since closed down, the records destroyed long before Bush ran for president. So I began searching for prominent rehab centers of the era, wondering if I could luck upon a former employee with knowledge who would be willing to blow the whistle on something from the distant past. It was a long shot, to be sure, but one has to start somewhere.

While conducting Internet searches and reading articles, I inadvertently came across material raising questions about Bush's opponent, John Kerry. I was reaching dead ends on Bush, but some of the questions about Kerry were pretty easy to check out. They had to do with his Vietnam War record and whether he had exaggerated or misrepresented his hero status.

Long story short, I was able to obtain the citations for Kerry's Purple Hearts, which are given when a soldier is injured during combat. I was also able to obtain the records describing the war event that led to each injury and merited a medal. From what I could see, the dates and injuries didn't match up quite right. One of the narratives made no mention of Kerry getting injured. I asked some military experts, and they told me that a Purple Heart always includes a narrative describing the event where the injury occurred. Taking a more careful look at the records, which were provided by the Pentagon, I saw that some of them were not originals from the Vietnam War era. For example, one document recounting Kerry's actions in the war was signed by a Navy secretary who had served long after the Vietnam War: Admiral John Lehman. *Why?*

I got one explanation from some Pentagon contacts. They told me that after Kerry returned from the Vietnam War, he supposedly threw his medals over the White House fence in protest. This supposedly accounted for why he did not have the original paperwork for his medals. Therefore, said the Pentagon officials, he later needed to apply for duplicates, which were then written up and issued long after the fact. It didn't make a lot of sense to me. Were we to understand that Kerry threw documents about his medals—pieces of paper—over the White House fence along with his medals?

I figured the military might have some outstanding records that would explain the apparent gaps. I called a colleague, CBS Pentagon correspondent David Martin, and asked if the Pentagon had more Kerry records to fill in some holes. Martin put in a query and got back to me. The Pentagon told him that the records we already had were the only ones available regarding Kerry's Vietnam War–era service.

So, although I was coming up empty on Bush, the Kerry information merited further research. At that point, I approached the folks in the CBS political unit, based in Washington, DC, to tell them what I was looking into.

I barely got the first few sentences out of my mouth before the political unit producers began scoffing.

"Vietnam was a long time ago," one of them remarked.

"The records just probably aren't very reliable," another offered.

"Maybe," I said. "But we should at least try to find out if there is a story here." And, playing my version of the Substitution Game, long before I called it that, I pointed out to them, "If Bush's records didn't seem to match up, we wouldn't just say, 'Vietnam was a long time ago.' We would look into it."

Not only were they disinterested, I sensed hostility. At that point, I realized that there would be no support for the idea of digging into anything involving Kerry—only Bush. I was so disgusted that I walked back into my office and dumped my whole Kerry file into the trash.

I came to put a lot of thought into experiences like that during my two decades at CBS News. I began to focus a lot of energy on making sure I opened my eyes to see what was really going on around me rather than wearing blinders and missing the true story. And I came to believe that it was crucial for me to report on stories and views that were off the typical narrative.

It was not always easy, but I did have a lot of success in that arena. I was often encouraged and supported by some top-notch bosses and colleagues. I was honored to be part of CBS News teams covering many stories in the CBS tradition of "fair and fearless." Some of the stories received award recognition from my peers and are part of the public record.

But there are many untold stories. And they reveal a lot about the death of the news as we knew it.

The Untold Stories

All reporters have stories that are "killed" from time to time, sometimes for good reason. Most often, it is because there simply is no time and space for all of them in the broadcast. Difficult choices are made every day regarding which ones are sidelined.

But there was a noticeable shift over the years in how these decisions are made, as groups seeking to promulgate narratives got better at manipulating us through an intricate system of pressures and incentives—everything from social media attacks to contacting CBS corporate executives to rewarding those who published narratives with "exclusive" tips and interviews. In the end, I'm sorry to say, because of this shift, some of the best investigations I ever conducted never saw the light of day on the CBS flagship broadcasts. Other times, stories were intentionally shaped by New York producers or managers in a way to obscure the original meaning and blunt their impact.

Many of my colleagues at CBS Washington complained about this trend, which seemed to reach a crescendo in 2011. It accelerated when there were a number of management changes on *CBS Evening News* and among the executive ranks. There were days when some producers in the Washington, DC, bureau became so angry about all of it that they physically threw objects. Some correspondents became so frustrated, they stormed out of the office and threatened to quit. But they always came back. *We all have bills to pay, and where else would we go?*

In some respects, I was the canary in the coal mine. These hits on journalism were being felt at news outlets beyond CBS, but they impacted me sooner and more deeply than some other correspondents because I was assigned to investigative reporting. By definition, nearly everything I touched could be seen as potentially dangerous to somebody's important narrative. As powerful interests learned which PR companies and law firms to hire, whom at CBS to contact or pressure, how to exploit social media, and how to enlist a network of "nonprofits," websites, and quasi-journalists to help them, it became clear that those of us on the receiving end were ill prepared to fight them and protect our independence.

Several years before I quit CBS, I began to recognize these organized efforts for what they were. I spoke with the CBS attorneys in New York, whom I enlisted to review my investigative stories prior to going on air.

"The people who are trying to stop our stories have figured it all out, and we always end up playing defense," I told them. "They have nothing but time and money to spend twisting us up in knots to respond to spurious complaints. We—and I mean all of us in journalism—need to address this and develop our own strategies so we aren't always playing defense and can do our job."

They did not disagree. But there simply was no will at the network level to organize a strategy to protect our news space from these sorts of attacks and influence. I also suggested that journalism conferences hold workshop sessions on this topic. I believed it to be at least as important as dozens of other challenges we were examining. But it was a cry in the dark. Had we taken proactive steps as an industry back then, I believe we might have been able to put the brakes on The Narrative getting its grip on journalism and destroying our reputation along with it.

Meantime at CBS, the pace of stories getting killed quickened, even as I was breaking more news than ever. There were full-fledged internal power struggles between managers who wanted to keep doing good journalism and those who clearly wished to slant the news for any number of reasons.

I came to learn the same struggles were under way at other news organizations. Colleagues at other networks and print outlets whispered of similar stories. I heard secondhand accounts from newsmakers— they would tell me that certain national news reporters had their stories blocked up the chain. I started to hear reporters complain that particular topics were becoming banned or off-limits. Members of Congress and their staff reported a similar phenomenon on Capitol Hill, where hearings on untouchable topics were banned; investigations on certain special interests were forbidden.

At one investigative news conference, I sat down for drinks with one of my counterparts at a competing television news network.

"How are you still getting your stuff on TV?" he asks, looking at me over a cocktail.

"Well, actually, most of it doesn't get on TV," I confessed.

"They make me put mine on the web instead of TV," he told me.

"That's exactly what they're doing to me," I tell him.

Meantime, some managers at CBS were directing my time and effort away from the groundbreaking investigative reporting I'd done for years. It felt as though they wanted to distract me or keep me busy chasing mundane stories that most anyone could do—stories that weren't original or investigative and frankly required few journalistic skills. It was frustrating to see the shelving of original investigative stories uncovering corruption and fraud using inside, on-the-record whistleblowers. For instance, I was assigned to do a meaningless live shot at Reagan National Airport outside Washington, DC, for "breaking news" about a chute that inadvertently deployed on a small airplane somewhere else in the country.

After several years of failing to defeat this trend, I realized it was a losing battle. That prompted me to walk away from my job at CBS—a job where I'd thought I would work until the day I chose to retire. Even the reporting about my departure was dominated by false narratives. And I haven't written before about what really happened in detail. More on that later. Here is a sampling of a few of my most memorable stories that died a tragic death at CBS. And, contrary to The Narrative circulated about why I left CBS, these reports were not all killed for reasons of liberal narratives. The story is far more complicated than that.

Earmarks

In 2007, *CBS Evening News* assigned me to cover the practice of Congress's "earmarking" taxpayer money for various projects and interests, circumventing the normal budget checks and approvals. By all accounts, the segment became among the most popular features produced for *CBS Evening News*. I regularly exposed a wide array of outrageous projects, equally distributing the shame among Democrats and Republicans, whether the malfeasance involved waste of tax money or criminal behavior.

My reporting on earmarking gained so much attention that some in Congress credited—or blamed—me for the eventual movement to do away with the practice of earmarking altogether.

Meantime, behind the scenes, CBS was feeling pressure over the stories. I felt the enthusiasm for them among management decline. For example, when I first got the assignment, my New York bosses agreed it would be important to highlight all sizes of earmarks from large amounts to small. Not only do the "small" earmarks of a few hundred thousand dollars apiece add up, they're often more relatable examples than the big earmarks amounting to tens of millions of dollars. One good example of a small earmark that people related to was $500,000 in taxpayer money earmarked for a teapot museum in North Carolina after lobbyists convinced Representative Virginia Foxx and Senator Richard Burr, both Republicans, that it was a bright idea.

But as powerful interests began to apply pressure on CBS, the internal philosophy about what we should and should not report changed. One manager told me I should focus only on earmarks of at least a million dollars. Not long after that, another manager raised the bar. He said we should probably report only on earmarks that were "tens of millions" of dollars. And then, as popular as the feature had been among our viewers, it faded away.

The Swine Flu "Epidemic"

In October 2009, I put the finishing touches on what today remains one of my most important and eye-opening investigations.

Using hard data, I discovered that there was a relatively negligible number of H1N1 swine flu cases in circulation, despite alarming claims by the Centers for Disease Control (CDC) that we were being overrun by illnesses and deaths.

Initially, the top brass in the news division marveled at the findings in my report. I had spent weeks obtaining lab test results from all fifty states and Washington, DC, because the CDC had refused to provide the information when I asked for it. One upper-level CBS news executive excitedly told me that it was the only original story on swine flu that he had seen anybody produce!

But for reasons unspecified, an influential senior producer intervened to keep the story from airing. She said that maybe we could report the information "when the whole thing is over" as part of a "look back." I was baffled. So instead of the story airing on *CBS Evening News*, I published it online. Later, when hundreds of people filed for compensation for injuries and deaths linked to the swine flu vaccine, I couldn't help but wonder if some lives might have been saved if we had reported the true prevalence of swine flu on our newscast. Some people might have declined the experimental, hastily developed vaccine if they had known that the risk of contracting swine flu was so low.

New York Welfare Scandal

Also in 2009, I got a tip about a run on ATM machines by welfare recipients in New York. People were standing in long lines to make cash withdrawals. There were so many people that the machines were running out of cash. There was a simultaneous run on beer and lottery tickets at convenience stores. Walmart also reported welfare recipients buying big-screen TVs and other luxury items using their welfare cards. Store clerks and managers were confused as to whether those were allowed purchases.

What was going on?

It turns out what prompted the whole mess was an action by New York governor David Paterson. He had accepted $35 million from the left-wing activist George Soros, which fulfilled a federal matching requirement so New York could get another $140 million in federal taxpayer "stimulus" money for low-income families. All the cash was directly deposited into the accounts of welfare recipients for them to buy school supplies for their children. The grants amounted to $200 per child. However, New York officials did not announce the grants in advance or notify recipients what the money was intended to be used for. As a result, many who saw the money deposited into their welfare accounts did not buy school supplies.

"People were taking money straight from the ATM machine to buy beer, cigarettes, and lottery tickets," one convenience store operator told me.

What's worse, even documented drug abusers received the un-restricted cash. Social workers worried that some of them had gone straight to buy illegal drugs to feed their habit.

"We were seeing people with drug abuse problems getting a thousand dollars in their bank account," a social worker told me.

I got the green light to do a story on the debacle for *CBS Evening News*. I was able to collect video of the lines of people at ATM machines, I got interviews with convenience store employees who were the first to discover what had gone wrong, and I got an interview with a New York State social worker. I asked for an interview with the governor, which was not granted. For additional balance, I located a welfare family that actually spent the money on what it was intended for: school supplies.

I put the story together, and it was approved up the chain. But when the edited version of the video was sent to New York for screening, there was a hiccup. A managing producer called me in the Washington, DC, bureau and said, "We can't run the story."

"Why not?" I asked, dumbfounded.

"Because all of the people are—of a certain color," he replied. The welfare recipients who had been photographed standing in line were black.

"But it's not a subjective comment on anything; that's just what happened," I countered.

"I know," he said. "I just can't take the hit for that right now."

I pointed out that *all* of the main players in the story were black, not only the ones who had misspent the money but also the family that spent the money properly, the concerned social worker, and some of the convenience store employees who were criticizing what had happened.

"We aren't villainizing anybody by race," I pointed out.

"I know," he said. "I just can't do it right now."

Other media did cover the story, but we did not air our report on *CBS Evening News*.

Feed the Children

On February 9, 2010, I reported a landmark investigation showing that the charity Feed the Children had greatly exaggerated how much aid it was providing to victims of the catastrophic earthquake in Haiti. Feed the Children's website originally claimed the charity was playing a major role at a refugee camp. It stated that it had been chosen by UN agencies "to provide food and milk for the entire camp" of thousands of hurricane refugees. I sent one of our cameras and a producer on the ground to the camp in question to check it out. I had already investigated alleged dysfunction and fraud at the charity.

Our CBS crew found one Feed the Children employee on-site at the refugee camp. Shockingly, she told our camera and producer that the charity had not fed anyone and had not gotten the contract to provide food and milk for the camp, after all. This was in direct contrast to Feed the Children's explicit claims at home. We aired an explosive story on *CBS Evening News*.

After my report aired, I got new leads from numerous Feed the Children whistleblowers about other alleged instances of mismanagement and fraud at the agency. It wasn't easy, but my producer and I managed to convince a dozen Feed the Children employees to speak out on camera. Obviously, they were risking their jobs in doing so, but they believed it was important to expose what they had seen firsthand. There were so many who were willing to talk that we interviewed them in a group on tiered bleachers.

Feed the Children was already in meltdown mode from our first report and its ongoing inner struggles. Now the agency became aware that I was working on a potentially devastating part two with interviews with numerous insiders. Somehow the charity's representatives contacted CBS and pulled strings. The second news report never aired. It was awful for me to have to report back to the brave whistleblowers who had risked their jobs that their stories would not be told. They would nonetheless suffer the consequences of stepping forward, because Feed the Children management would surely obtain the names of those who had spoken with us.

These sorts of events are disastrous to the credibility of news organizations. Word gets around that some powerful interest was able to kill a news story on CBS. Whistleblowers then no longer trust those of us who ask them for information. It is a crippling blow to the cause of good journalism.

People to People

On July 12, 2011, I conducted another lengthy and difficult investigation. This one was on a different nonprofit called People to People. The organization was accused of tricking high school students into believing they had been specially chosen as "student ambassadors" to represent the United States on trips to exotic countries. But I learned that People to People was little more than a high-priced travel agency that exploited student lists to get customers. The group even forged a legislator's signature on student invites! It wasn't easy, but I even got an on-camera interview with the legislator who confirmed that her signature had been forged. Some of the People to

People trips were allegedly so poorly managed that they resulted in student injuries and deaths. I also learned that People to People had settled numerous lawsuits filed by families of victims, sworn them to secrecy, and allegedly threatened them with dire consequences if they spoke of what had happened to their loved ones.

A watered-down version of my investigation ultimately aired on the CBS morning program. What viewers did not know was that I had originally produced the story for *CBS Evening News*. It was supposed to air the first week that Scott Pelley became the anchor of the broadcast in 2011. It had the approval of the CBS lawyers whom I had asked to review my stories. It had a go-ahead from producers up the chain. But shortly before the day it was scheduled for air, it fell off the calendar in New York without notice or explanation. Pretty soon, Pelley contacted me with changes and softened the story. One of the main things he cut out was the information about some students who died on the trips. The process of Pelley modifying my story continued for several weeks. Every time I made his suggested changes, chipping away a little more at the heart of the story each time, Pelley came back with still more changes. It was altered so much that it became a mere shadow of its former self. Finally, as if he did not remember having asked me to remove mention of the students' deaths in the first place, he asked me, "Did anybody die on these trips?"

"Yes," I told him.

"Well, maybe you should put that in the script," he said.

Is this real life?

I came to realize I was riding on a never-ending carousel. The merry-go-round was a way to kill a story without saying so, while maintaining the charade that the story might actually air.

I came up with a nickname for this sort of treatment, which became sickeningly common: "death by a thousand cuts." Those who were working to shape narratives at CBS in New York, often for reasons known only above my pay grade, would not outright refuse to air a story; quite the opposite. They would say, *"It's a terrific story, world class!*" They would pretend to be looking for the perfect time to air it. They would act as if it just needed a little phrase cut out here and there. Then they would delay and delay until I understood that the story was never really going to air.

Unions and Green Energy

Another of the more interesting stories I investigated at CBS got killed, in my opinion, for the sin of being politically inscrutable.

It began in April 2012 when a CBS investigative producer brought me a tip. She said labor unions in Michigan were angry that a lot of tax money under President Obama's green energy stimulus program—a combined $300 million—was actually being given to Korean companies and Korean workers, and used to buy Korean supplies at US-based plants.

We managed to get photographs and video of Korean employees of the US plants doing hands-on work that was supposed to be done by Americans. We secured a rare interview with a union leader, usually loath to criticize a Democratic administration, describing how American workers were upset about their jobs being taken by foreign workers. He said Democrats in Congress, the Labor Department, and President Obama had ignored or rebuffed union inquiries about all of this.

Initially, all of my colleagues involved in reviewing the story I wrote said it was fascinating. But it hit a brick wall with the *CBS Evening News* executive producer in New York. She reviewed the script with my producer, and my producer called me with bad news.

"She hates it," the producer told me.

"What doesn't she like?" I asked. "Can we rewrite it and fix it?"

I had grown accustomed to this executive producer changing scripts and story direction for her own politically motivated reasons.

For example, for years at CBS, I researched stories about congressional spending's waste of tax money. But once this particular woman became executive producer, when she came across the phrase "tax money" in my stories, she changed it to "federal money." She didn't explain why. My producer and I figured that since she was highly partisan in her work, inflicting her overtly liberal ideals into the broadcast, she likely did not want viewers reminded that the government could be misspending their tax money. So it was "federal money."

There were times I could fix a story to take out the parts that collided with her worldview. I hoped that would be the case with the report about stimulus funds spent on Korean workers. The story was too good to kill.

What were the executive producer's objections to the story? "She wasn't specific," my producer told me. "She just hated the whole thing."

Later, I spoke on the phone with the executive producer. First, she did not want us to use the term "foreign" workers. I asked what she would call them: "Non-US-citizen workers, perhaps?" she kind of sputtered. Instead of giving an answer, she made it clear she simply did not want to air the story at all. Her explanation was that it "lacked outrage." Her reasoning was that "Some people *did* get jobs with the stimulus money—even if they were Koreans. So there's no outrage."

I argued that the stimulus money had been intended to help US companies and workers. But more important, it wasn't my goal to "outrage" anybody. Viewers would be free to form their own conclusions about the story. Some might agree with her that there was no problem with stimulus money going to foreign workers. Others might agree with the unions. I said that she shouldn't block the report simply because people might draw a different conclusion from the one she wanted them to draw.

It was one of the first times I remember becoming keenly aware of news managers who believe it is their role to keep viewers from learning information if it does not lead them to the "right" conclusion. The story did not air on *CBS Evening News*.

School Lunch Fraud

A month later, in May 2012, I reported on a nonpolitical story (which actually describes the majority of my reporting). It was a report for *CBS This Morning* about school lunch fraud. My producer and I had learned exclusive information about a major investigation into big companies that provide the food that goes onto the lunch trays in our public schools. Some of the companies were being charged with cheating schools and taxpayers out of millions of dollars. The morning show producers enthusiastically accepted our story idea and pushed us to quickly travel and shoot it. As I rushed to get them a script, it suddenly hit a brick wall. Literally overnight, the producers decided they did not want the story. They didn't even want to read the script.

My producer and I spent an afternoon trying to solve the mystery of why the turnabout. During this time frame, we would spend hours in a week speculating as to what reasons the politically conflicted New York managers would find to stifle our original stories. And it was not only politics we found ourselves navigating. Maybe a story I was working on involved an investigation into someone tied to a corporate partner of CBS. Or an advertiser. But I could not for the life of me understand whose toes we would be stepping on with a report about school lunch fraud. So I did an Internet search of the phrase "school lunch." The apparent answer to my question immediately popped up: the search returned stories about a new initiative by First Lady Michelle Obama to improve school lunches. My producer and I concluded that someone affiliated with *CBS This Morning* had pulled our story because they thought it could somehow be seen as a negative reflection on Mrs. Obama's efforts.

Dreamliner Fires

In January 2013, CBS assigned me to investigate the Boeing 787 Dreamliner fires. After extensive research, I uncovered some exclusive information. I got video of a massive fire several years earlier caused by a faulty prototype of the Dreamliner battery. It burned down a whole lab and injured a key worker who had tried to blow the whistle on the design flaws. I convinced this whistleblower to do an on-camera interview. I even got an interview with a former top federal safety official who said that the information I'd uncovered amounted to a "smoking gun."

Both CBS producers working on the story with me said it was an incredibly strong piece. I put it through my normal legal review, and the lawyers gave it the green light. And the story was approved by the senior producer in Washington, DC.

But in New York, it hit a dead end. First, incredibly, the executive producer of *CBS Evening News* told me she did not understand why the report included video of the battery fire. She wanted the video removed. The request was so absurd that I recognized it as the start of the old "death by a thousand cuts" routine. I knew it was a waste of breath for me to argue for the story. There were reasons—perhaps ones I would never know—that the executive producer did not want the story to air.

The discussion with the executive producer took place over the phone with her in New York and me in the Washington, DC, newsroom. Also listening on the extension on my end was my DC senior producer, who had approved the story. I made eye contact with him while the executive producer in New York ranted, and I silently signaled to him with my finger making a cut motion across my throat, meaning it was pointless to go on.

Next, I offered the story to the CBS Saturday morning news. When one broadcast turns away a story, it is often possible to "shop it around" to another one that has different tastes or is less conflicted about the subject at hand. The executive producer of the weekend morning program viewed the finished piece and said he would be delighted to have it. I would fly up to New York at the end of the workweek to appear on the set and introduce the story.

But Friday afternoon, before I flew to New York, I got a call from the Saturday morning executive producer. He told me that he was sick about it, but a certain CBS executive had stepped in and basically put a hold on my Dreamliner investigation. The only reason we could think of for these attempts to block the story was pressure from Boeing. I had been told by multiple people I'd contacted in researching the story that Boeing, an extremely influential and politically well-connected company, had been using its influence to stop news stories about the Dreamliner fires and prevent congressional hearings.

My Dreamliner investigation never aired. The incident became one of the final straws that led me to depart CBS ahead of my contract term.

In 2019, there were two deadly crashes of Boeing 737 Max aircraft. They killed a combined 346 people. The reporting surrounding those tragedies included some of the very same themes I had uncovered in my reporting about the Dreamliner six years earlier—in the story that did not air. Internal Boeing company emails about the 737 Max showed Boeing employees referring to cover-ups and criticizing what they said were flaws in its design and processes. I couldn't help but wonder: *If my Dreamliner investigation exposing related alleged weaknesses had aired, could it have helped prevent the 737 Max tragedies?*

Ted Cruz

Around the same time CBS management killed the Dreamliner story in early 2013, *CBS Weekend News* assigned me to do a story on the new US senator from Texas, Republican Ted Cruz. Even before he was elected, Cruz was a target in the national news media because of his conservative views.

To get the interview with him, I laid the usual groundwork. Through multiple calls and contacts, I had to convince Cruz staffers that I was not setting out to do a hit job. I just wanted to do an informational story about him as a new, potentially influential member of the Senate. I was used to working hard to get an interview, particularly when it came to Republicans. Many in the GOP distrust the media, view reporters like me as liberal, and seemed particularly suspicious of CBS News at times.

Cruz agreed to the interview, and I produced a fair profile that told viewers where he stood on a variety of issues relevant to the upcoming congressional session. The weekend news executive producer read and approved the story. But it was never scheduled for air. It died on the vine. I was never given a reason, but my own theory is that the story did not air because it did not portray Cruz as a fire-breathing villain.

Nakoula Nakoula

Later in September 2013, with the help of my intrepid producer, Kim Skeen, I secured an agreement for an interview with the maker of the short film "Innocence of Muslims." The Obama administration had falsely blamed the film for triggering "spontaneous protests" at a US compound in Benghazi, Libya, on September 11, 2012, that got out of hand and resulted in the death of four Americans. In fact, we learned, Islamic extremist terrorists had executed a planned attack. It was not a protest, spontaneous, or prompted by the film. A House investigation concluded that US officials had been given plenty of warning that a terrorist attack was imminent and the State Department had denied US diplomats on the ground the additional security they repeatedly requested. Documents later revealed that Obama officials, including then secretary of state Hillary Clinton, worked hard to cover up after the fact and furthered The Narrative that it was all the fault of "Innocence of Muslims" and its producer, an Egyptian Christian named Nakoula Nakoula. Nakoula was then arrested for probation violations related to earlier bank fraud and identity theft charges.

Now, a year later, Nakoula was about to be released from a halfway house in California. I spoke with him on the phone and arranged to interview him in the car as he was whisked away to a safe house under threat of death. I had no idea exactly what he would say, but it was an interview most any journalist would have wanted. We had worked for months to get this exclusive opportunity.

By now, as I have described, I had become familiar with the increasingly frequent pattern of *CBS Evening News* management, and sometimes other executives, killing important investigative and original stories. This was a big one. I knew that if it were to air, I would need the man at the top of the news division to get involved and issue explicit guidance to the troublesome *CBS Evening News* executive producer. Therefore, I put in a call to CBS News president David Rhodes. I told him about the interview we had scored. I said I would need his help in getting it on the air.

His response gave me a sick pit in my stomach. "Isn't that—old news?" he asked.

My heart fell. I had no idea why he did not want the interview and would be left to guess.

These are just a few examples of what I would estimate to be approximately a hundred legitimate stories blocked from airing, not for reasons of time and space but having to do with narratives, political or corporate influence, or special interests. As the incidents grew more frequent and the universe of "acceptable" stories grew ever narrower, I concluded there was little meaningful journalism for me to do in the current news environment.

My Last CBS Award Season

My final years at CBS felt dysfunctional, to say the least. Even as it became more difficult for me to get some of my best stories onto the news, I did manage to get a few great pieces onto the air, and they comprised some of the strongest reporting of my career.

How was this possible?

Keep in mind that CBS was not a monolithic organization; many people there did support investigative and original reporting. In between my stories getting blocked, managers were assigning me to investigate new ones, such as the Benghazi, Libya, terrorist attacks and the Dreamliner fires. Often, CBS managers complimented and cheered on my stories. It just became a game of wondering which ones would get blocked when, by whom, and for what reasons.

In 2013, for the first time since I began doing investigative reporting at CBS News, the New York managers skipped over me when it came to picking stories to consider for the Emmy Awards. Normally each year, they sent me and other correspondents a note asking which stories from the previous year we considered worthy of entry. My producer and I figured that our exclusion this time was for one of two reasons: it was either a snub by certain managers to make us feel excluded or an attempt by those managers to make sure my stories that were off narrative were not rewarded on the public stage.

As I and my producer, Kim, discussed the Emmy snub, Kim had an idea: instead of entering the Emmy Awards through CBS, we could submit the entries on our own. It just meant I would have to cover the fee myself: $250 per entry. Normally, we might have recommended one or two of our stories for CBS to enter. But Kim and I felt we had three strong candidates, and since we were entering ourselves, we decided to go ahead and submit them all! One was a group of investigative stories holding Congress accountable. It included an undercover investigation into fund-raising by Republican freshmen. A second group of stories was my reporting on the cover-ups and security issues behind the September 11, 2012, Benghazi attacks. I had broken a lot of international news on that front. The third entry was a group of stories investigating waste, fraud, and abuse involving government spending on green energy initiatives. We submitted all three entries in March 2013.

The Emmy nominations were announced in July, and I was pretty surprised. All three stories received Emmy nominations! I had already received a Daytime Emmy Award earlier in the year as part of the *CBS Sunday Morning* team entry for Outstanding Morning Program for my report "Washington Lobbying: K Street Behind Closed Doors." I had also recently received finalist recognition from the prestigious Gerald Loeb Awards for a series of stories I called "Taxpayer Beware."

When the final 2013 Emmy Awards were announced in the fall, my group of congressional stories won the award for Outstanding Investigative Journalism. I was also honored to be invited to present at the Emmys that year. It was my strongest year ever in terms of recognition by independent journalism award groups.

It would be my last full year at the network.

The dysfunction at CBS, particularly surrounding my position there, was punctuated after the 2013 Emmy recognition. It had always been stated policy at CBS that anyone who entered the Emmys at his own expense would be reimbursed for any entry that received a nomination. But when I submitted to get reimbursed for my three Emmy nominations, CBS gave me the runaround. Eventually, the New York official handling reimbursements told me that the president of the news division, Rhodes, now said there was a new policy: that CBS would no longer pay back employees who received only nominations; we would be reimbursed only if we actually won the competition. Months later, one of my entries did win, and I again applied for reimbursement. This time, I was told the CBS policy had changed again and reimbursement still would not be provided.

All of that added up to a strange atmosphere that gave me an escalating sense that I needed to get out, even though my employment contract required me to stay. I didn't know it at the time, but my decision to leave would be sealed in an unexpected way.

The Long Good-bye

All jobs have their ups and downs. The work I did at CBS was extremely high pressure. But throughout the years since I first got into journalism in college back in 1980, there weren't many days when I didn't love what I do.

Yet the changes in the industry took their toll. I began worrying frequently about what I saw as ethical lapses and improper influences at work. I often brought these concerns home, discussing them with my husband and daughter, Sarah, then in high school. Sarah had not previously seen or heard me be so distressed about the job I'd always loved. One day, while listening to me describe a problem at work, she looked at me, shrugged, and asked, "Why don't you just quit?"

The idea of leaving CBS had never crossed my mind until I heard the suggestion come from my young daughter's lips. It suddenly dawned on me that I had a way out. I did not have to be miserable, left sitting on the sidelines to watch what I felt was the degradation of a once great news division. The problem was, I had recently signed a new contract. So I came up with a plan. If I were willing to walk away from any promised severance package and the prospect of a job at another news organization, there was nothing CBS could do to make me stay and finish the contract.

First, I had to work out the plan with my husband, who was not happy to hear me suggest walking away from a very lucrative contract. But once he agreed, I told my producer, Kim. She understood. I did not give a heads-up to my agent, Richard Leibner, because I did not want to put him into a difficult position. At the time, he represented a number of CBS on-air talent and managers, and I didn't want him to have to keep a secret as I planned my exit. I also did not want him to try to talk me into staying. I spent a couple of weeks moving my personal belongings out of the office, carrying a box or two out to my car every few days. Then, one Friday afternoon, I left for what I intended to be the last time. I contacted Richard and told him to call CBS and inform the people there that I would not be back on Monday.

Maybe I was naive to think walking away could be that simple.

All hell broke loose.

There were phone calls, meetings, and consternation from New York to Washington, DC. After some discussion, Richard told me that CBS was not going to let me out of my contract. I responded, *"Well, I'm not going to come to work anymore, so they can just fire me."* Everyone wanted to know why I was quitting. I had decided not to give reasons, because I thought it would give the false impression that things could be fixed, and I knew they could not. To me, the problem wasn't just the executive producer or anchor or even CBS; the whole industry was changing. I had heard others inside CBS voice complaints similar to mine. I heard them from my colleagues at other networks and at national print publications. The main difference between me and them was that I could walk away.

Ultimately, Richard told me CBS was not going to let me out of my contract until I did an exit interview. He urged me to show up for a meeting with Washington Bureau Chief Chris Isham. I did so. At that meeting, I got blindsided. CBS business affairs executive and lawyer Chris Andaya had flown in from New York and was in Isham's office when I entered for the meeting. I immediately asked to get Richard on the phone. Isham and Andaya said it wasn't necessary, that this meeting was "nothing negative." I reached for the landline telephone and tried anyway but could not get through to Richard.

Both Isham and Andaya said that everybody in management had been blindsided by the news that I was unhappy and wanted to leave CBS. Andaya reiterated "how much everyone at CBS respects you and values you and your work." They asked if there was "anything, anything at all that CBS could do" to get me to change my mind. I said there was not.

Andaya also wondered why I "had not shared [my] unhappiness with management before now so that we could try to address the issues and fix them so you would stay." Isham added that he didn't understand why I hadn't shared the issues with him so that he could work on solving them for me.

I said I had communicated many issues to management over time but I felt that things had continually gotten worse. I thanked Isham for having been my advocate over the years but reminded him that he had often been unable to help, telling me at times that "they [New York] just don't want investigative reporting." Isham said he had absolutely no idea that the problems were so serious that I would want to end my contract early. I reminded him that after CBS killed the Dreamliner story, I told him I couldn't foresee finishing out my contract in this environment. He said he remembered but hadn't taken my comment too seriously. Andaya said that although he knew I had made some complaints about different issues over time, I had not initiated a big, single conversation with management during which I tied it all together, said I wished to quit, and gave them the chance to fix things. He repeated that "CBS values you very much" and "you're well respected."

I replied that I'm not the type to "run around with my hair on fire" when there are problems but that I had expressed my serious concerns clearly to a lot of people over time. I said I was confident that most people in management were well aware of the issues. Andaya asked why I did not want to provide a list of what was wrong so they could fix it. I said there was no point in hashing all of that over because it made no difference at this stage. These were big, systemic things that couldn't be fixed. I said I just wanted to leave on good terms, that I didn't want to take any money and they could use the salary they would have been paying me to hire three or four young journalists who would do the sort of work that CBS New York wanted. I told them it was a win-win.

Next they brought up workplace harassment incidents that had occurred and asked if those cases were why I was leaving. They reminded me that they had addressed the incidents properly. I told them that my desire to leave had nothing to do with that. They seemed concerned that I might be considering legal action over those events, and I reassured them that I did not intend to sue.

"I just want to leave," I said.

Next they changed tack. They accused me of giving my Dreamliner story to Fox News after CBS had killed it. They said someone at Fox News supposedly told Rhodes that that was the case. (Rhodes formerly worked at Fox News.) Of course, it was an utter fabrication. At the time, I had no friends or contacts at Fox News. My contacts had been limited to a couple of appearances on Bill O'Reilly's program, *The O'Reilly Factor*, to discuss some of my investigative stories, and all of those had been arranged by Rhodes. I wondered why Rhodes would fabricate a tale that I had given the Dreamliner story to someone at Fox.

I would have none of it. I told Isham and Andaya I wanted to call my lawyer. They seemed to become worried and assured me they were not accusing me of anything, but I insisted they had. They tried to tamp down the tension. Andaya repeated that their goal was to keep me working at CBS, reminded me that the company had signed a new contract with me not long before, and said he saw my stories on *CBS Evening News* quite often. Maybe, he suggested, I should not be so disappointed by how many did not air because my other stories did air. I told both men that I'd had twenty years at CBS and felt I had contributed a lot and got a lot from it. I said that I'd had more successes than failures, I enjoyed a lot of my time there, but that the last period of time had gotten too hard and there was nothing left for me to do. Andaya asked whether I understood that I'd signed a contract with CBS that lasted through a certain date. I said that I understood but had nonetheless decided to leave. The meeting ended, we shook hands, and Andaya was friendly. He said it was too bad we had gotten together under the circumstances of my wanting to sever my contract, but if there was absolutely nothing CBS could do to turn things around, he would go back to New York and see what he could do to help me end the contract early. I was hopeful.

Over the course of the next few weeks, as I continued to try to negotiate my way out of my contract, I also demanded that CBS investigate the false accusation that I had given the Dreamliner story to Fox. CBS told me there was no need to do an investigation because it agreed that Rhodes's supposed information turned out to be incorrect. I said I still wanted it looked into so that we could learn where the false claim had originated. Meantime, the longer the process dragged on, the more aggravated I became. At one point, Richard and my attorney told me that Rhodes said he would allow me to sever my contract only if I would agree that every time I mentioned CBS in the future, in any context for the rest of my life, I would pay the network $100,000. Ridiculous.

Then one day, Jeff Fager asked me to fly up to New York to meet with him. At the time, Fager was chairman of CBS News, above news division president Rhodes. When I arrived at Fager's office at 524 West 57th Street in New York, Rhodes was there, too. Fager did the talking. He was friendly and conciliatory. He told me he was aware that numerous producers and correspondents were having internal issues with certain CBS management, particularly surrounding *CBS Evening News* after big personnel changes had been made there. He said that a number of CBS veterans who usually never complained had done so and that he planned to fix things. He asked me to give him some time to do that.

"How long?" I asked.

Six or seven months, he said. If after that I didn't think some of the big issues had been resolved, he would consider letting me out of my contract early.

Although I believed that the problems I had with CBS were beyond repair, I agreed to cool my heels. By this time, I had been trying to end my contract for weeks to no avail. In the end, I actually gave CBS almost one more year of my time. There were more ups and downs; more story successes but continued challenges that I felt posed an untenable ethical situation for me to operate within. We negotiated an out and parted amicably, considering the craziness. Fager emailed me on my last day, March 10, 2014, "*I have always enjoyed working with you, and I will miss you. You are a very talented reporter, and I know you will do well no matter what comes next.*"

I felt as though a weight had been lifted off my shoulders. And I didn't expect to work in the news business again. After all, the industry had largely changed around me into something that felt foreign to the journalism I practiced. Besides, who would want to hire someone like me who digs up stories that make the wrong people uncomfortable?

Almost immediately, CBS insiders who were hostile to my reporting, and the special interests who had worked hard to stop my stories, crafted a false narrative about my departure from CBS. Some claimed I had been fired over my alleged conservative bias, a topic that had never come up in my departure discussions. Others claimed I quit over CBS's liberal bias—also a topic not raised in my departure discussions. Of course, once these anonymous claims and theories got planted, published, and amplified by news websites, blogs, and social media, the false version of what happened is the one that dominated. Propagandists at the usual suspects—Vox, Salon, Media Matters, *Mother Jones*, Wikipedia—claim false narratives were true, though they were never rooted in fact.

In any event, as fate would have it, my father was diagnosed with brain cancer as I negotiated my final departure from CBS in 2014. I was able to use the time immediately after I left my job to help my mother provide hospice care for him at home. During that period, I received a surprising number of job nibbles. I was not ready to commit to anything right away and certainly wanted to avoid jumping into another bad situation. Over the course of the next year, I did some freelance writing and broadcast news reporting. And then in 2015, I accepted a position created at Sinclair Broadcast Group to do a weekly television news program. Sinclair's head of news, Scott Livingston, who had worked with my producer Kim decades before as a photographer in Baltimore, wanted to feature the kind of accountability reporting I used to do on CBS: stories about whistleblowers and watchdogs; off-narrative stories; no spin. It was the genesis of my Sunday program, *Full Measure*. I'm lucky. I don't know of many old-school national journalists who feel satisfied with their current job situation. Not all that many feel they can conduct good journalism unfettered by undue outside influences.

With that in mind, let's continue our journey with an examination of the broader challenges posed by today's information landscape. What makes getting at the simple truth so very difficult?

CHAPTER 2

The Narrative by Proxy

Never has there been a US political figure more adept at directing The Narrative than President Donald J. Trump. By pressing the "Send" button on a 280-character tweet or turning a phrase at a press conference, he's able to send news media around the world into a tailspin as they dissect, fact-check, and criticize his remarks. Even the criticism can serve Trump's purpose. We end up talking about the issues he puts before us.

The national journalists I interviewed agree that "the news" in the era of Trump has reached a crisis state. "Reporting is supposed to be gathering the facts, making sound judgments and putting it together with knowledge, not laying it out there without context," a former top network news executive told me. "But today, there's no context. Or it's manipulated." Like others, she blames Trump. "Our president is leading it. He's setting narratives and manipulating facts. And our culture has become deaf to it because there's distrust of everyone, everywhere. Nobody knows what to believe anymore."

But as much as Trump has taught the world about the power of narratives, I find the narratives promulgated by the press to be more problematic. After all, most politicians push narratives of one kind or another. Their careers may depend on how well they're able to convince us to believe the narratives that tell their story in the most positive light. Heck, regular folks are not much different! Most of us have a story line we want to tell about ourselves. Maybe we want to explain what we're all about. Whether we are politicians or ordinary folk, we drive home selected information that lines up with what we want others to believe about us. It is a bit like pursuing our own personal PR campaigns, whether we realize we are doing it or not. And it's a large part of being human today, especially in the age of the Internet and social media, which provide a 24/7 outlet to express any given narrative about ourselves or others.

But journalists aren't supposed to do that. *Our* goal should be to resist blindly reporting narratives. We must critically look at other facts and views to ensure that we are on point with independent news gathering. This means sorting through the multitude of narratives put before the public on any given day or setting them aside entirely to reveal the deeper story about who wants us to think what—and why.

It is our job as reporters to do the legwork and identify topics of public interest on which to report rather than copy one another or rely on what's whispered into our ears by special interests. These days, my favorite stories to cover are the ones that no powerful interests have brought to the table. No PR firm is pitching them. No paid pundits are pushing them. No lobbyists or think tanks are trying to convince Congress to talk about them. Too often, we in the media simply choose among the narratives laid out before us or advance our own. Believe it or not, some journalism schools are even teaching the next generation of reporters to do just that. Young minds are being taught that using bias in news reporting is a virtue! That it is their job!

One Pulitzer Prize–winning journalist who worked at a national nonprofit newsgroup told me about an exchange of internal emails within the organization a few years back discussing this very issue. The way the story goes, some of the elder journalists on the email thread were criticizing the new and growing trend of reporters mixing their own opinions into news stories without labeling them as such. This journalist said he was shocked to find a formidable contingent of fellow journalists defending the practice. They argued that it is perfectly acceptable for journalists to put their opinions in their reporting "as long as their opinions are based on facts."

To old-school journalists like me, this mentality is unthinkable. But it has become disturbingly commonplace.

Don't get me wrong; there are still hundreds of accomplished reporters producing excellent journalism every day. Some are mentioned within these pages. But they are operating within a landscape that is foreign to some of us who have been in the professional news world longer than a decade. A new breed of reporter is dominant at many news organizations: the kind who think it is their job to convince you to believe whatever they personally believe; the kind who don't look for original stories, seek out research, or open their minds to opposing views. They are the kind that spin the news according to what they want you to think. They ignore facts that contradict their story line. They get their ideas from other reporters, quasi–news media, PR firms, political operatives, and talking points pushed out by special interests. In other words, their sources are those in the business of pushing narratives. They justify their one-sided position by citing propaganda-laden rhetoric, such as "We don't report both sides when it comes to the earth being round or flat, after all."

All of this makes it more difficult for truly independent-minded journalists to do their jobs. The good work gets lost in a sea of clickable, predictable, biased, conflict-oriented reporting. And today, good journalists can find themselves reporting to managers who are more interested in supporting a particular viewpoint than getting at the facts.

A top national news executive I interviewed for this book told me, "I believe most reporters' mistakes happen when someone made up their mind [about a story] and set out to prove it happened. To prove a thesis. Any facts they learn that are mitigating, they throw away. That's not news journalism."

Since mid-2015, the dominant press narrative has been decidedly anti-Trump, so much so that if an independent reporter remains objective and takes no particular personal position on Trump, that reporter is accused of being pro-Trump. The same thing was not true, in my experience, with President Barack Obama. Those of us who wrote stories remaining neutral on Obama, did not report on Obama in particular, or wrote stories that included positive mentions about him were not routinely stalked and criticized by their peers or on social media for being "pro-Obama." But mention Trump in any context other than an attack, and some people treat it as if it were a cardinal sin. It can take a lot for a reporter to stand up to that kind of pressure and remain true to the facts.

This concept of reporters finding themselves accused of bias if they aren't overtly anti-Trump is worth raising at the outset because this book contains substantial discussion of the media's treatment of President Trump. Some will try to spin critiques of the media herein as advocating for Trump. In today's Alice in Wonderland environment, reporters who show overt bias against Trump view themselves as fair. But reporters who are fair to Trump are labeled as biased.

University of Texas history professor Alberto Martínez learned that lesson firsthand. A liberal supporter of Democrat-Socialist Bernie Sanders in 2016, Martínez also happens to be a factual analyst of news and information. He's certainly no Trump advocate. Yet he told me he finds the media's false narratives about The Donald to be shocking in audacity.

In 2019, Martínez published a book titled *The Media versus the Apprentice: The Devil Mr. Trump.* In it, he analyzes twenty-one infamous news stories about candidate Trump. He tells me, "In every instance I found that reporters and pundits grossly exaggerated and distorted the facts, sometimes by one hundred percent and other times by much more." In one example, the media claimed that Trump had originally advocated for a wall along the entire two-thousand-mile southern border but then flip-flopped and said that we don't need a wall where there are natural barriers. Actually, the media got it wrong; Trump had never wavered. As Martínez notes, "[Trump] always proposed a thousand miles all along," having pointed to the fact that there is no need for a wall where there are natural barriers on the border as early as 2015.

"The media narrative about candidate Trump was that he was The Villain: both recklessly moronic and unapologetically evil. News pundits claimed that Donald Trump, above all, insulted minorities and women," Martínez tells me. "But did Trump really insult women and minorities more than he insulted anyone else? No. Like him or not, Trump is an equal-opportunity offender not motivated by race, birthplace, gender, or sexual orientation." Martínez reviewed the evidence and found that Trump reserved some of his strongest insults for "rich, white men." That included calling Republican Karl Rove "sick," "a loser," "failed," "a dummy," "dopey," "an establishment dope," "a total fool," "a moron," "a biased dope," "a total loser," and "an irrelevant clown." Martínez also found that Trump had attacked Republican senator Rand Paul as "failed," "a fool," "lowly," "a lightweight," "didn't get the right gene," "truly weird," and "a spoiled brat without a properly functioning brain."

"No news articles admitted that Trump flung some of his strongest insults at prominent white men—because it didn't fit the narrative," Martínez concludes.

The Devolution of "the News"

In the past few years, we have experienced a sea change in terms of how the media do their job. The emergence of Trump as a viable political candidate accelerated this devolution. No longer do reporters keep their opinions firewalled from their stories; their stories are rife with their opinions. Dubious anonymous sources are repeatedly relied upon even after they have proven shamefully unreliable. Basic fact-checks go unconducted as long as the "news" furthers an anti-Trump narrative. Egregious reporting mistakes are made by the same outlets—and sometimes the same reporters—over and over again yet are magically forgiven. Journalism ethical standards are bent or suspended so that Trump can be covered more aggressively and in a one-sided fashion, because the media have declared him to be uniquely dangerous.

Some of the best evidence for this can be found in a June 2017 opinion piece by Mitchell Stephens published in Politico. Stephens is a journalism professor at New York University. In his article, he unabashedly cheers on the end of media objectivity under a new president attacked by the media. The headline: "Goodbye Nonpartisan Journalism. And Good Riddance. Disinterested reporting is overrated."

It is shocking that a journalism professor, someone teaching up-and-coming professionals in the field of news, considers disinterested reporting to be "overrated." Disinterested reporting was once thought of as a pillar of good journalism. Tossing that aside is a bit like a medical school professor telling future doctors that diet and exercise are overrated.

Stephens goes on, in his very partisan piece, to heap praise upon the *New York Times*' public descent into blatant partisanship:

> *Our most respected mainstream journalism organizations are beginning to recognize the failings of nonpartisanship—its tepidness, its blind spots, its omissions, its evasions. It was news when the patriarch of American journalism, the* New York Times, *finally used the word "lie," in a headline on atop its front page on September 17, 2016, to describe a Trump assertion.*

Stephens notes with satisfaction that other media followed the *Times*' lead:

> *Other legacy journalism organizations began more regularly calling out Trump's "falsehoods," if not actually accusing him of lying. About a week later, the* Los Angeles Times *declared, also on page one: "Never in modern presidential politics has a major candidate made false statements as routinely as Trump has." . . .*

> *On April 5, 2017, the* Times, *reflecting the new order, quickly changed a headline online from "Trump Says Susan Rice May Have Committed a Crime," to "Trump, Citing No Evidence, Suggest [sic] Susan Rice Committed Crime." Pelley at CBS upped the ante with "divorced from reality."*

What Stephens sees as worthy of acclaim, I view as irresponsible. A journalism professor is teaching a new generation of reporters to inject agendas and opinions into their news stories; to forget about the firewall we used to attempt to maintain between news and opinion. The kind of reporting Stephens applauds would have gotten a traditional news journalist fired not all that long ago. Today, it's part of what makes it so easy for The Narrative to take hold and can make it so difficult for the truth to be told.

There is endless evidence of this sort of slanted reporting today. For example, based on the following headline in *The Atlantic*, who would you guess came out on top in a September 2019 election in North Carolina—Democrats or Republicans?

North Carolina Gives Republicans a Wake-Up Call
The results of a special election portend trouble for the GOP in 2020 . . .

It sure sounds as though Republicans got their clocks cleaned. After all, the headline states that they got a "wake-up call" and the results "portend trouble for the GOP." So you might be surprised to learn, as I was, that Republicans actually had a successful night. It's laid out right there in the article—if you get past the misleading headline. In a congressional race, the Republican candidate beat the Democrat by a "far wider" margin of 4,000 votes. Separately, in a state special election, "Republican Greg Murphy won, as expected."

How on earth did those Republican victories elicit a headline implying they lost? It is as if somebody was bent on pushing a particular narrative regardless of the actual election results.

In this way, we can see how those pushing The Narrative don't care much about facts. Pesky facts that contradict a narrative are nothing more than a nuisance to brush off. Reporters simply devise creative ways to dispense with them.

Another example demonstrates nearly every trademark of a narrative and the perils that come with advancing it blindly. On April 25, 2020, Politico reports that President Trump owed the Bank of China tens of millions of dollars in a loan coming due in 2022, as he dealt with China on the coronavirus pandemic. The implication is that Trump could not possibly be as tough as he needs to be on Chinese leaders for unleashing Covid-19 and covering up its seriousness because he is beholden to them.

The "news" makes headlines around the world. "Trump Owes Tens of Millions to the Bank of China—and the Loan Is Due Soon" blares the headline on Politico. "Donald Trump's Debt to China" reads *The New Yorker*'s headline. "Trump Owes Millions to Bank of China for Building Loan, Records Show" screams *National Review*.

But it isn't true.

Shortly after the story circles the earth, the Bank of China issues a statement saying it had held the Trump loan for only twenty-two days before selling it to a US real estate firm in 2012. In other words, the tens of millions of dollars Trump supposedly owed China—due soon—was owed for only three weeks back in 2012; eight years before.

Obviously, the facts negate the whole idea behind the story. But the media are not about to admit making a mistake. Politico changes its headline and the details of the story but does not issue a formal correction or apology for three more days. Other media pretends the false information has little bearing on the issue. *National Review* simply changes "Trump Owes" China to "Trump Owed" China. Past tense.

Politico could have avoided the error if it hadn't failed to follow a basic rule taught to nineteen- and twenty-year-old journalism students: contact those mentioned in a news story to ask for comment. Of the three reporters bylined on the article, Marc Caputo, Meridith McGraw, and Anita Kumar, none apparently thought to contact the Bank of China prior to publication. And apparently, no editor thought it was necessary to do so.

Had they done their job, they likely would have learned prior to publication that there was no current loan to Trump from China, thus avoiding their major flub.

SUBSTITUTION GAME: Dubious claims that fit The Narrative are advanced, while supported claims that are off narrative get buried. The same month, March 2017, that the media reported Trump had "mishandled classified information," the Justice Department inspector general (IG) found that ex–FBI director James Comey actually *had* mishandled classified information regarding President Trump. The IG even recommended that Comey, an Obama appointee, be charged with a crime. But officials at the Department of Justice declined to prosecute, saying they didn't think Comey had meant any harm. By any neutral assessment, the finding that Comey mishandled classified information for political use against Trump should have been a huge story; but it wasn't. Yet the unsupported and challenged allegations against Trump? Global headlines. A cursory Google search for mentions of Comey's alleged crime [Comey, mishandled, classified] turns up 95,000 results. And relatively few of the top results are news reports by mainstream national news organizations. Compare that to the treatment Trump got: a similar search [Trump, Russia spy, classified] returned 3.3 million results—critical news reports by the top media outlets and blogs around the globe.

The "Two Sides" Fallacy

Open-minded news consumers tell themselves that they can avoid falling victim to The Narrative by making sure they listen to pundits with opposing views or sample a diverse range of publications, such as Fox News, CNN, the *Washington Times*, and the *Washington Post*. This is a common fallacy.

The problem is, even when you seek different viewpoints, the opinions often surround the same two or three topics. You are still being

fed a steady diet of The Narrative. The fact that so many news organizations fill time and space highlighting the same stories points to
a narrative's success. It means that operatives pushing narratives are
getting certain stories put front and center and managing to keep
competing stories hidden from public view. We have made it easy
for these operatives to accomplish their goals by inviting so many
of them to fill TV news airtime. Cable news is saturated with their
opinions and commentary.

Their dominion over America's news landscape is no accident. An
entire cottage industry is made up of special interests that recruit,
train, and supply the news with camera-ready media commentators.
They are carefully schooled on how to deflect inconvenient facts; how
to flip answers to uncomfortable questions into key talking points;
how to interrupt someone who is giving an opposing viewpoint so
that it cannot be heard; when to arch an eyebrow, offer a smile, or
pound the desk. There is a science behind how these emissaries get
themselves placed in front of the cameras at prominent news organizations. Nearly every newsperson I interviewed for this book cited
the prevalence of political pundits, panels, and analysts as among
the biggest problems with news today.

"In the first two decades at CNN, I can't think of a time when there
were nine people on the set offering political opinions or analysis
about anything," says Ralph Begleiter, a former CNN colleague of
mine, commenting about today's trends. "Nine people was considered an uninformatively large number of people to have on the set at
one time, so it just didn't happen. Today, it's flipped. The opposite is
the case."

Begleiter is the Joe Friday of journalism, a just-the-facts guy. Starting in 1981 and during his two decades as CNN's world affairs correspondent, he reported on events from a hundred countries and
seven continents without any spin. This just-the-facts guy sounds
frustrated when I ask him about the state of the news: "The more
people you can cram on the set at one moment to express themselves
in very short sound bites, and the most sharp-pointed, then you've

scored big in the current environment. But it's superficial because they don't get to talk for more than a few seconds at a time."

Beyond the inherent superficiality, there are other problems. Yes, experts and analysts can sometimes add meaningful context and perspective. But by allowing so much of the news to become an unfettered platform for so many political operatives, we in the media have transformed ourselves into propaganda tools. There is simply no debating whether this is the case. *Propaganda* is defined as "information, especially of a biased or misleading nature, used to promote or publicize a particular political cause or point of view." It is, by definition, what political operatives produce. When we constantly supply them with microphones and give them space in our news stories, we're serving their narratives. We are pretending they offer true news value even as our viewers and readers know that isn't the case. Consumers realize they're getting predictable, prepared, slanted talking points.

All of this flies in the face of what news operations ought to be doing. In journalism college at the University of Florida many years ago, I learned that it is perfectly fine to listen to stories and angles pitched by pundits, commentators, and PR professionals. But when you determine there may be something newsworthy in their information, your job isn't over; it has just begun. *They've* told you what they want the public to know. They've given their spin. You, as the journalist, are supposed to figure out the facts. What is the rest of the story?

Looking at matters that way, it is mind-blowing to consider what news organizations often pay these political operatives to distribute their talking points on our media outlets! A part-time consultant, analyst, or contributor to a news network can easily earn anywhere from $60,000 to $120,000 a year to provide commentary. If anything, *they* should be the ones paying the media in this dysfunctional relationship. After all, they get the benefit of the bargain; they're getting mass distribution channels for their messaging. But somehow the script is turned: *we* pay *them* for the privilege of giving voice to their propaganda. The more polished and partisan their presentation, the more in demand they are, because that's seen as "good" television.

On top of that, too often we fail to adequately disclose a particular consultant's conflict of interest. In November 2013, former CIA acting director Michael Morell was hired by a PR strategy firm, Beacon Global Strategies, populated by Hillary Clinton loyalists. Two months later, as if executing a PR strategy, Morell simultaneously got hired as a contributor at CBS News. When he appeared on camera at CBS, opining on matters leading up to Clinton's second run for president in 2016, CBS did not disclose to viewers that he was working for the Clinton-connected PR group. As the presidential election drew closer, Morell temporarily resigned from CBS, endorsed Clinton in the *New York Times*, and helped with her campaign, then promptly rejoined CBS after Clinton lost. (Morell has said in public interviews that he always acted independently as a commentator and not to advance a political cause.)

Political consultant Doug Schoen is a fixture on Fox News, at times commenting on issues related to political controversies about Russia and Ukraine. Though he often adds an interesting perspective, I have not seen a disclosure to viewers about the fact that he has made a fortune as a foreign agent representing the pro-Russian Ukrainian billionaire Victor Pinchuk. There is nothing legally wrong with the relationship; it is just that viewers should be provided that context so that when Schoen gives his opinions, they know who is helping pay his bills.

Like the former CNN mainstay Ralph Begleiter, many old-school field reporters and producers despise the "Brady Bunch" roundtables dominating so much of today's news. Good journalists develop their own original ideas through solid reporting. But they find little appetite for originality among news managers kept busy arranging the never-ending carousel of appearances by political operatives spewing talking points.

"If [some of today's news managers] have to choose between general news and something like a Donald Trump tweet or attacks by Mitch McConnell or Nancy Pelosi, they choose the personality route," one prominent national news executive tells me. He describes himself as "progressive" and has worked at the highest levels in both broadcast and cable news. "Today, the media chooses to be partisan," he says. "You turn on a station, and if you wiped out the graphics, you'd say to yourself, 'This is a Democratic or Republican Party TV station.' There are very few shows that are down the middle journalistically in any way or make any attempt."

I blame this trend, in part, for the dramatic polarization of the news. With political consultants at the ready, virtually all news discussions are reduced to Right vs. Left. They are there to tell you how today's news fits into the existing Narrative. This stokes deep divisions among the viewing public and makes it nearly impossible to report factually on important issues. Every news event automatically becomes politicized because when there's breaking news, political pundits are available and, by God, there's time to fill!

If there is a major drug bust, it is reduced to a debate over which party's policies are at fault for allowing drugs to get out of control. Economic trends? They are boiled down to disputes over whether Obama or Trump gets credit for the good or blame for the bad.

Even an event as inherently apolitical as a hurricane becomes a political argument over global warming and which politicians failed to prepare for it. If President Trump announces plans for a Space Force, it turns into bickering over whether his ideas are silly or Obama decimated NASA.

In August 2019, a man shot six Philadelphia police officers. While the scene was still hot, political candidates and pundits were already live on the news and social media, weighing in with political blame. All of the opinions were expected, none enlightening.

One of the most instructive examples came with the coronavirus outbreak in early 2020. On WMAL news radio in the Washington, DC, area, Larry O'Connor interviewed me about media coverage. He pointed out that no sooner had cable news anchors announced the health emergency than they had "swiveled their chairs" and started a political discussion with Democratic and Republican pundits about the whole mess, which immediately framed everything in political terms.

"You almost never see a story. They do panels," says another former top network news executive who spoke with me. "I find that hard to take."

Responsible news media free from the influence of narratives would resist being used like this. They would reject playing a scripted role in creating an atmosphere of disunity.

These are some of the reasons we have fallen in the eyes of the public. A poll by Scott Rasmussen in the fall of 2019 shows just how far. Seventy-eight percent of Americans said political reporters promote their own personal agenda using news events as props for their narratives, rather than seeking to accurately record what really happened. A meager 14 percent of voters said national political reporters actually do their job and report what's happening.

Worst of all, we are allowing ourselves to be used as tools of destruction.

Media Self-Censorship

One frightening way in which the news becomes slanted is through self-censorship.

It was the fall of 2016, just before the presidential election, when I first noticed this befuddling and dangerous movement getting a serious grip on our media landscape. News media, social media companies, politicians, and government began inserting themselves as never before into the role of determining what you should and should not learn about. They argued that *they* needed to decide which facts and opinions you should hear and which you should be forbidden from knowing about. They did it—they said—for your own good, lest you lose track of The Narrative or form the wrong conclusions with your own brain and all. They assumed this Big Brother role at the urging of less-than-independent third parties such as political groups, nonprofits, corporate interests, academic organizations, and journalism associations.

Handpicked curators and fact-checkers began staking claim to what they deemed to be the truth. They insisted they have a unique ability to divine ultimate truth—even when the truth is impossible to know, such as what will happen in the future, or a matter of debate, such as whether a certain policy is good or bad.

They'll tell you what to think. It cannot be left up to you.

A young generation of Americans may never know any differently. They may not remember a time when the Internet was a free, unfettered resource where information was available at everyone's fingertips without unwanted intervention from "curators" and self-proclaimed "fact-checkers." They will not know that "the news"—at least when we were at our best—was a place where different sides of stories were told, where reporters did not pretend to know answers to unanswerable questions, where journalists didn't provide theories about what newsmakers might be thinking, where news organizations did not restrict the parameters of who is to be believed, what words can be used, and which facts and views get banned. They will have no recollection of journalists fairly questioning the powerful no matter their political persuasion. They will be unfamiliar with a time when reporters strived to keep their personal opinions sidelined from news pieces. The idea of a news report that does not tell viewers exactly what to think—will be foreign.

Somewhere along the way, we seem to have become comfortable with the idea—in fact, we have come to *demand*—that third parties intercede to help us avert our eyes from that which someone decides we should not see.

Some people argue that it does not qualify as "censorship" when news organizations and social media giants wipe selected facts, stories, and opinions from our view. These people say it qualifies as censorship only when the government does it. But I think the changing media landscape has expanded the definition of censorship. It now encompasses behavior by nongovernment entities, as well. No longer are there clear, bright lines between news, politics, advertisers, and corporations—if there ever really were any. For example, a social media company may accede to political demands to control public access to certain information in order to earn protection from government regulations, taxes, or penalties. News organizations may report stories a certain way to please advertisers and corporate partners, who donate to political parties.

One demonstration of the censorship cross-pollination comes in January 2020. The Association of American Physicians and Surgeons (AAPS), an alternative to the American Medical Association, sues Congressman Adam Schiff, a California Democrat. The group accuses Schiff of abusing government power and infringing on free-speech rights by censoring vaccine safety information online.

"Who appointed Congressman Adam Schiff as Censor-in-Chief?" asks AAPS in announcing its lawsuit.

The dispute over the control of vaccine information is part of an ongoing propaganda war waged by pharmaceutical interests using health officials and surrogates in the media. Their goal is to eliminate and discredit certain vaccine safety information, reporting, and scientific studies—such as those advanced by AAPS—that could end up hurting the financial bottom line of vaccine makers.

According to the lawsuit, in February and March 2019, Schiff urged Google, Facebook, and Amazon to deplatform or discredit what he claimed was inaccurate antivaccine information. AAPS said the information was not antivaccine; it was pro–vaccine safety—and accurate. "Within 24 hours of Schiff's letter to Amazon dated March 1, 2019, Amazon removed the popular videos *Vaxxed* and *Shoot 'Em Up: the Truth About Vaccines* from its platform for streaming videos, depriving members of the public of convenient access," the lawsuit says.

The lawsuit claims Twitter also made online modifications to control access to vaccine safety information. In May 2019, Twitter announced plans to insert a pro-government disclaimer above Internet search results leading to an AAPS vaccine-related article. The government disclaimer stated, "Know the Facts. To make sure you get the best information on vaccination, resources are available from the US Department of Health and Human Services." AAPS takes issue with the implication that its information or any other material not on government websites is less than credible.

Facebook got in on the alleged censorship, too. An Internet search for one particular vaccine safety article by AAPS is directed, instead, to government websites such as the World Health Organization (WHO), the National Institutes of Health (NIH), and the Centers for Disease Control (CDC).

"The Internet is supposed to provide free access to information to people of different opinions," states AAPS executive director Dr. Jane Orient. "AAPS is not 'anti-vaccine,' but rather supports informed consent, based on an understanding of the full range of medical, legal, and economic considerations relevant to vaccination and any other medical intervention, which inevitably involves risks as well as benefits."

AAPS argues that Schiff overstepped his bounds: "Under the First Amendment, Americans have the right to hear all sides of every issue and to make their own judgments about those issues without government interference or limitations. Content-based restrictions on speech are presumptively unconstitutional, and courts analyze such restrictions under strict scrutiny."

When it comes to media coverage of the lawsuit against Schiff filed by AAPS, vaccine industry propaganda again rules the day. A writer named Olga Khazan at the left-leaning *Atlantic* oddly claims that an Internet ad for a book dedicated to "children who had to suffer due to adverse vaccine reactions" is hard-and-fast proof of the antivaccine dangers lurking on the World Wide Web. She goes on to criticize AAPS by calling it a group with "a belief that mainstream science isn't always trustworthy."

It would be funny if it were not so serious.

Khazan, *The Atlantic*, and others who act as propagandists are implying that we must always view "mainstream science" as trustworthy and never question it—despite thousands of reasons why good journalists should be far more circumspect. After all, mainstream science once gave its stamp of approval for pregnant women to use thalidomide, later blamed for birth defects in forty-six countries. Mainstream doctors once said smoking was good for us. Mainstream scientists gave approval to the RotaShield vaccine to prevent diarrhea in newborns, but the medicine was found to cause a fatal disorder in some babies. You can probably think of more examples. The point is that although mainstream medicine has done incredible things to save and improve lives, it is hardly infallible or without conflicts of interest. What is seen as mainstream science today is often not mainstream science tomorrow. Yet many in the media today suggest that acknowledging this fact is heretical. Journalists used to believe part of their job was to question in a rational way. Now many see their job as convincing the public not to question certain narratives and bullying those who do.

Falsely casting vaccine safety reporting as "antivaccine" is a powerful narrative that has become more successful as online control and manipulation has grown stronger. I had no idea how influential those forces were until I was assigned to cover the topic of vaccine safety at CBS News in the early 2000s. First, the network asked me to investigate vaccine injuries among military troops. There was long-standing controversy over illnesses caused by anthrax vaccine, and questions about plans to start inoculating troops and then the general public with smallpox vaccine after the Islamic extremist terrorist attacks of September 11, 2001. Smallpox vaccination in the United States had been suspended in 1972 because the disease was deemed eradicated. But after 9/11, there were new fears that terrorists would weaponize the virus in an attack on the United States. I covered the restart of the smallpox vaccine program, which was ultimately shelved due to deaths among some of the early recipients.

Back then, many national news reporters were covering vaccine and prescription drug safety issues. We got the expected pushback from the pharmaceutical industry, but it didn't intimidate us. None of us was labeled a "conspiracy theorist" or "antivaccine" for reporting factually on these important topics of interest to so many Americans. My own reporting received national recognition from independent journalism groups such as the Emmys and Investigative Reporters and Editors, and my work was cited favorably by a Johns Hopkins University neurologist in *The New England Journal of Medicine.*

But as the pharmaceutical industry began to feel the squeeze from widespread national news coverage, it sent its lobbyists to flex their muscles on Capitol Hill. Hearings on prescription drug and vaccine issues were successfully slanted or blocked. The industry also found it could sway news divisions with its billions of advertising dollars. The result is what I now consider to be one of the most pervasive and successful false narratives of our time: the idea that there are no vaccine safety issues, that all vaccines are to be accepted unquestioningly by all members of the public, that all links between vaccine injuries and autism have been debunked, and that all scientists or journalists who report on vaccine safety are "antivaccine, tinfoil-hat conspiracy theorists." There is a meaningful difference between the positions that "vaccines do not cause *most* autism cases" and "vaccines have *never* caused a case of autism." Books have been written on this phenomenon, but I will demonstrate it in practice with one key example.

One of the more recent news stories I broke on this topic was in January 2019. It was a stunner. It would have been an even bigger story if not for The Narrative and the media's self-censorship.

By way of background, there's a special arrangement between Congress and vaccine manufacturers: when children or adults allege injuries from vaccinations, they cannot sue the vaccine makers; they have to sue the US government in a special federal vaccine court. In this court, the Department of Justice takes the side of the vaccine manufacturers, defending the vaccines. If the victim wins, it isn't the vaccine manufacturers that have to pay damages; the money comes from a trust fund funded by a tax we all pay on every dose of vaccine we receive.

To summarize my story in 2019: I had just obtained a copy of a shocking and important affidavit recently written by a top medical expert. Not just any medical expert—he's the man the Department of Justice used as its expert witness in federal vaccine court a decade earlier to fight vaccine-autism injury cases: the world-renowned pediatric neurologist Dr. Andrew Zimmerman. Dr. Zimmerman could be counted on to firmly testify that vaccines cannot cause autism. There is no more pro-vaccine or credible authority than Dr. Zimmerman.

That's exactly why Dr. Zimmerman's new affidavit is so momentous. In 2019, he goes on record to say that vaccines *can* cause autism, after all. He says he first reached that conclusion in 2007 based on advances in science and his own experience with a patient.

But there's more.

Dr. Zimmerman's affidavit goes on to say he told Justice Department attorneys back in 2007 that he believes vaccines could cause autism. Coming from an authority like him, it was news of epic proportions. If word of this had gotten out at the time, it stood to upend everything the government had long claimed in its attempt to debunk vaccine safety concerns. And it would have validated what some parents and scientists had insisted for years: that there is a vaccine-autism link, after all. But instead of letting the news out, Dr. Zimmerman says, the Justice Department covered it up, fired him as an expert witness, and misrepresented his medical conclusions in court—as if he'd said there is no way vaccines can cause autism.

I report all of this news on my Sunday television program, *Full Measure*, in 2019. I fold in exclusive interviews with former members of Congress and their staff, both Democrats and Republicans, who detail how they had been blocked over the years from investigating vaccine safety and autism links due to the pharmaceutical industry's hold on Congress. My story is fact based, well sourced, and free of conjecture or opinion on my part. Viewers are left free to form their own conclusions about a complex topic.

Nonetheless, The Narrative kicks in. As my story is circulated on the Internet, it is flagged by Facebook as untrue. I found that out when viewers sent me screenshots showing what happened when they tried to share the *Full Measure* article: Facebook added a label claiming my report contained "false information," as judged by Facebook's nameless "science" fact-checkers.

There are two possibilities, both equally objectionable. Either Facebook's fact-checkers are propagandists for the vaccine industry, or they are so ill informed that they simply do not know the facts. This has implications far beyond the story at hand and should worry all of us.

It's one more reason why I felt a chill down my spine in January 2020. That's when the World Health Organization (WHO) and Google announced a partnership to make sure the public got steered only to government-approved information online about the coronavirus outbreak that reportedly originated in China. As reported by The Verge website, "Searching Google for 'coronavirus' will now send users to a curated search results page with resources from the World Health Organization, safety tips, and news updates."

There's that word again: "curated."

I first heard the term applied to controlling news and information in October 2016 when President Obama introduced the concept at an appearance at the private research university Carnegie Mellon. Obama claimed a "curating" function had become necessary. The public at large had not been asking for any such thing. Instead, it was the invention of powerful interests that apparently felt the need to get a grip on public opinion—interests that were losing the information war online. But the concept is contrary to the nature of a free society and an open Internet. It would take some clever manipulation to convince the public to allow such "curating."

"We're going to have to rebuild, within this Wild, Wild West of information flow, some sort of curating function that people agree to," said Obama. "... [T]here has to be, I think, some sort of way in which we can sort through information that passes some basic truthiness tests and those that we have to discard because they just don't have any basis in anything that's actually happening in the world."

As far as I know, that signaled the start of what would become a global media initiative to have third parties insert themselves as arbiters of facts, opinions, and truth in the news and online.

There would be little wrong with this government-led nudging to "curate" news if the following were the case:

IF THE GOVERNMENT'S INFORMATION WERE ALWAYS CORRECT.

IF THE GOVERNMENT ALWAYS TOLD THE TRUTH.

IF THE GOVERNMENT ALWAYS KNOWS THE WHOLE TRUTH AND ALL THE FACTS AT ANY GIVEN MOMENT.

IF THE GOVERNMENT AND ITS OFFICIALS NEVER HAVE ULTERIOR MOTIVES OR CONFLICTS OF INTEREST.

Of course, we know that each of those conditions is ludicrous. So exactly how did Facebook become convinced to manipulate online information?

The effort can be traced back to the conservative-turned-liberal smear artist David Brock, the pro-Clinton, pro-Obama mastermind behind the propaganda group Media Matters for America and its network of nonprofits, websites, and political groups. In early 2017, Brock was caught bragging to donors about supposedly being responsible for pressing Facebook into curating material. "We've been engaging with Facebook leadership behind the scenes to share our expertise and offer input on developing meaningful solutions," read a Media Matters briefing book at a retreat in Florida.

When I look at curation or censorship of news and social media today, I instinctively think of Media Matters. There are conservative groups that also strive to have that kind of influence. But I am unaware of any right-wing groups that have attained equal success at the game—certainly, none that have forged partnerships in secret meetings to convince Facebook to take on the role of censoring and shaping public information.

Even some Republicans who have made their careers as smear artists have found themselves outmatched in this new world order. The political consultant Roger Stone is one of them. CNN, MSNBC, and Fox News banned Stone from appearing on their networks in 2016 after a successful campaign by Media Matters to eliminate him from public view. Stone's enemies did not just wish to provide counterpoints to his rhetoric; they sought to limit him from the public discussion entirely, as if he did not exist. *Whoosh!* Down the memory hole. To be sure, Stone—a Trump supporter convicted of lying to Congress in 2019—is a controversial figure. But his words and slurs are no more outlandish than those of many other bizarre personalities who have remained very much welcome in the media landscape. Some have made racist statements, provided false and defamatory information, been charged with or convicted of crimes, or committed ethical violations. But Stone alone was banned. As a result, he was denied a fair opportunity to answer the serious allegations lodged against him by the very news outlets that banned him. *You mustn't be allowed to draw your own conclusions*, the newsrooms decided. *We'll tell you one side of the story only. We'll tell you what to think about Roger Stone—and everything else.*

More media censorship was evident on October 11, 2019. NBC's *Meet the Press* did not want viewers to make up their own minds about something President Trump said at a rally about the son of former vice president Joe Biden. NBC tweets:

> The president held a campaign rally last night and attacked Hunter Biden. We cannot in good conscience show it to you.

The tweet then quotes NBC host Chuck Todd, chief curator:

> @chucktodd: "Politics ain't beanbag, but it isn't supposed to be this either. We all need to play a role in not rewarding this kind of politics"

This is a step beyond shaping information. It's withholding it entirely—with the excuse that it is for your own good. *The media know best.*

The more American and more honest approach, in my view, is not to ban any particular views, speech, or people except that which is illegal. The public should be left to make up its own mind. Facebook and Twitter would be far better off adopting the approach taken by the Twitter alternative Gab: simply state, as a matter of policy, that there are effective tools for users to filter content and block objectionable material. It should be left to users to sort through information. That way, news and Internet companies could avoid controversies such as whether they ought to be "fact-checking" political ads in which disputes are often matters of opinion, open to debate, or unknowable. Of course, that would also greatly diminish their stranglehold over The Narrative. And we seem to have traveled too far down the road in the other direction.

CHAPTER 3

Weaponizing The Narrative

The most insidious use of The Narrative is when it is weaponized to destroy. Nothing demonstrates that destructive power better than the chilling weaponization of #MeToo.

The #MeToo movement started around 2006 as a campaign to fight sexual harassment, misconduct, and assault against women, especially at work. It encompasses the idea that bad behavior, previously unrecognized or normalized, should be called out as the abuse that it is. It urges the media to spend more effort investigating the powerful men and the claims against them, instead of unearthing the romantic and sexual pasts of the female accusers, so that the women feel safe coming forward.

The trend reached a crescendo when #MeToo allegations touched dozens of high-profile figures in entertainment, news, and politics: media personalities Matt Lauer and Charlie Rose; TV network chiefs Roger Ailes at Fox and Les Moonves at CBS; *New York Times* reporter Glenn Thrush; singer Chris Brown; actors Bill Cosby and Kevin Spacey; movie producer Harvey Weinstein; political figures including representatives Pat Meehan, Blake Farenthold, John Conyers, and Trent Franks; Senator Al Franken; New York attorney general Eric Schneiderman—all casualties of #MeToo efforts to expose their alleged offenses.

Few people would argue with the notion that men or women who behave badly should be held accountable. For decades, too many misdeeds were tolerated. And when the behavior veers into criminal acts, there should be prosecutions. But as #MeToo became a global sensation, a dark and dangerous phenomenon developed on a parallel track: the #MeToo narrative was perverted into a tool to destroy. All a woman needs to do is to utter "Me too" or make an unsubstantiated claim, and some people insist that "she must be believed"— automatically and without question. *"Women would not lie about something like that"* goes the argument. *"Questioning a woman's claims or suggesting there should be proof before judgments only compounds the abuse the women suffered. It shows a level of un-wokeness that casts suspicion on the questioner."*

It is farcical to suggest women would never lie about abuse, rape, or harassment. The record is full of examples. In 1931, nine African American youngsters were falsely accused of raping two white women in Alabama. In 1987, a young black woman named Tawana Brawley falsely claimed she had been kidnapped and raped by a group of white men. In 2014, *Rolling Stone* published false rape claims made by a University of Virginia student that ultimately led to the reporter being found guilty of defamation. In 2019, a Florida woman accused a neighbor of rape. Her male friend killed the neighbor in anger. But the woman had lied; the neighbor was innocent.

But it is the Weinstein example that best sets the stage for what is to come. Long-whispered rumors of the famous film director's sexual misconduct—even criminal behavior—surfaces in the mainstream press in October 2017. The first accuser finds a warm reception in the media. More accusers speak out. The dynamic becomes a veritable feeding frenzy as reporters compete to become first to find the next accuser, another lurid story. The case against Weinstein feels as if it is open and shut. The women are believed. The reporters get accolades; Weinstein gets twenty-three years in prison.

It is in this charged atmosphere that dishonest people have discovered they can weaponize the #MeToo narrative.

You probably know about the high-profile accusations against then Supreme Court nominee Brett Kavanaugh. Democrats invited his main accuser, Christine Blasey Ford, to testify at his confirmation hearings in September 2018. Democrats and Republicans alike declared Ford to be a sympathetic and obviously traumatized figure. As high as the stakes were, senators seemed afraid to question Ford about her decades-old account, even though it was vague at best and evidence proved to be lacking.

Following what they viewed as the Weinstein recipe, which paid off for all concerned (except Weinstein), the media reported the most unsubstantiated and outrageous claims against Kavanaugh. The Narrative dictated that there had to be additional alleged victims—who were also to be believed. But in contrast to Weinstein, the case against Kavanaugh crumbled. Others who were supposedly present during his supposed attack on Ford claimed to have no memory of any such events, and Ford's own longtime friend cast doubt upon her claims. A second woman who also accused Kavanaugh of rape later admitted she'd never met him.

With the allegations against Kavanaugh looking thinner by the day, the whole saga quietly drifted off like the white strip of a jet contrail that's visible one moment, then nothing more than a shadow of something that might have been.

There are many lesser-known stories demonstrating how the #MeToo narrative has been perverted in today's slanted media environment.

Shades of Gray

At CBS News way back in the late 1990s when we were covering the sexual claims against President Clinton—and there were a lot of them—we passed over more accusations than we reported on the air. It's not that we didn't believe the women accusers; it's just that we did not have on-the-spot corroboration to elevate them beyond a case of "He says, she says." We believed that good journalism required something more concrete before we gave voice to such potentially damaging claims. The existence of multiple complaints, by itself, was not corroboration of uncorroborated claims.

Today, things are different. One uncorroborated claim may be viewed as weak—but if someone pushing a narrative can group together multiple people making equally weak claims, they are suddenly treated in a bundle as if they are all credible. The media concludes: *where there's smoke, there's fire.* But that is neither a logical nor a journalistic approach. And it offers bad actors an avenue with which to smear targets using false claims. "It's a numbers game," says one former television network news executive involved in a #MeToo scandal, though not directly accused of wrongdoing herself. "Reporters find multiple anonymous people saying the same unproven things and then treat them like they're all proven true."

Trevor FitzGibbon's story will give you chills—as it should. As a public relations professional, he says he was always aware of an old adage: *if you have two believable stories, you can take anybody down.* What he didn't know was that very thing would happen to him. He was targeted by multiple #MeToo accusations; the serious ones were later recanted. But not before they destroyed him.

It all started in December 2015. FitzGibbon was running his own progressive PR firm when he got a fateful call from his vice president, who told him, "In the past forty-eight hours, Human Resources has gotten six phone calls, all accusing you of sexual harassment."

"My heart kind of fell," FitzGibbon tells me in recounting the story.

To understand why FitzGibbon might have been targeted, it helps to know something about who his enemies included: some powerful liberal colleagues. First, he angered some of them by supporting Bernie Sanders over Hillary Clinton for president in 2016. Second, he represented a number of controversial clients connected to WikiLeaks, which became the enemy of powerful Democrats and establishment government figures after publishing damaging hacked documents about them during the presidential campaign. The clients included WikiLeaks and its founder, Julian Assange; Bradley Manning, who'd passed classified materials to WikiLeaks; Edward Snowden, the government whistleblower WikiLeaks once assisted; and Glenn Greenwald, the journalist to whom Snowden had leaked. The WikiLeaks connections will come into greater play in a moment.

Anyway, FitzGibbon says that before he even knew who his accusers were, their stories had somehow found their way into the national press. They must have had some pretty influential help! For most people with a story to tell, it is not easy to locate reporter contacts and get them to quickly publish an article on the case with almost no investigation. But FitzGibbon's accusers got immediate attention. The scandal reached national headline status at the speed of light—generally among the same group of publications that often sing from the same propaganda song sheets. They include Huffington Post, which "broke" the story, *The Guardian*, Vox, *The Nation*, the New York *Daily News*, the *Washington Post*, Medium, and Slate. It was clear: somebody was pushing a narrative against FitzGibbon and had well-placed media connections willing to lend a hand. FitzGibbon was far from a household name. Yet the slanted story against him was treated as if it were among the most important events going on in the whole wide world. Inexplicable—but for The Narrative.

"It's really interesting to see the Huffington Post [coverage]," Fitz-Gibbon tells me about the evolving early media coverage of the allegations. "Because at first, they say it was 'harassment.' A few hours later, it was 'assault.' And then that got spun into 'rape culture.'"

FitzGibbon says he was a victim of #MeToo's dangerous lack of nuance. He admits to inappropriate behavior toward women who worked for him or applied for jobs with his firm. But that's a long way from rape. Too often, the lines have become blurred. Bad or boorish behavior is conflated with crimes.

Before two weeks were over, FitzGibbon's entire staff had turned on him, and his company had to shut down. Three of his accusers enlisted help from the feminist lawyer Gloria Allred and filed criminal complaints against him. Most damaging of all: one of the women, civil justice attorney Jesselyn Radack, claimed FitzGibbon had "touched her breast" against her will, then raped her days later when she met up with him at a hotel for a rendezvous.

FitzGibbon admits that in hindsight, he was probably unintentionally inappropriate with female coworkers. But as for the criminal charge of rape, he tells an entirely different story. "It was one hundred percent consensual," he insists, referring to having sex with Radack.

The dastardly beauty of a false #MeToo narrative is that nobody wants to be accused of doubting a victim. But had reporters exercised basic journalistic practices and bothered to analyze the facts, they would have noted that Radack offered no tangible proof of her allegations. That didn't necessarily mean she was wrong, but FitzGibbon, on the other hand, was able to produce compelling evidence in his own defense: sexually explicit text messages and photos Radack sent him both before and after the alleged assaults. The photos show what appear to be her bare breasts, and her breasts covered only with see-through black lace. Among the accompanying text messages she sent: "You go first on fantasies. And I want a pic, too." In another message, Radack sent FitzGibbon what appears to be a snapshot of her bare bottom in thong underwear with a note, "although I don't think you're as interested in my bottom."

It took a year, but after reviewing the case, prosecutors "declined to file criminal charges" against FitzGibbon. Even then, he found himself hit with renewed, orchestrated attacks in the press. Seventy-two national organizations filed a public pledge in a media release vowing to never hire or work with him again, even though he was not going to be charged. That is when he began to suspect he was the target of an organized smear. "It was one of the first times that I realized that something else is at play," he tells me.

Proof of his suspicions seemed to be found within a sensitive PR document unearthed during that time period. It had been circulated among government contractors in 2010. The document described a wide-ranging strategy to combat "the WikiLeaks Threat" and to "sabotage or discredit" WikiLeaks supporters using "social media exploitation" and "disinformation." It included names of WikiLeaks associates—several of them FitzGibbon's clients—and placed their photographs in diagrams showing their relationship to WikiLeaks. Had FitzGibbon been targeted in a smear campaign because he'd angered high-ranking Democrats by representing WikiLeaks clients—and by choosing to back Bernie Sanders instead of Hillary Clinton? Interestingly, two targeted people named in the PR document were also discredited by sexual assault claims that—like FitzGibbon's—were widely publicized but never prosecuted: Julian Assange and Jacob Appelbaum.

The accusations against Assange involved two women who told a journalist matching sex stories. Each claimed she had been in the process of having consensual sex with Assange while he was in Sweden for a speech and that the sex had turned into rape. A rape investigation hung over Assange's head for seven years before it was finally dropped.

Meantime, anonymous accusers went public with sexual assault accusations against Appelbaum, a key WikiLeaks associate. Someone even started a website where Appelbaum's alleged victims posted tales of his supposed gropings and rape attacks. He was forced out of his job but—like Assange and FitzGibbon—never charged with any sexual crimes.

Is it just coincidence that three close associates of WikiLeaks were discredited by similarly unproven sex charges? Or was it part of a well-executed PR strategy that weaponized the #MeToo narrative?

A PR man by training, FitzGibbon worked hard to counter the narrative against him. But he found it excruciatingly difficult. Press reports covering the allegations were hopelessly slanted. He petitioned a lawyers' disciplinary body to punish Radack, an attorney, for her allegedly false accusations, but the group ultimately declined to punish her. In a letter, the law group said the question of whether Radack should be professionally disciplined for her actions "was close," but "the truth about what occurred in private is sometimes hard to prove." FitzGibbon then sued Radack for malicious prosecution and defamation. On May 3, 2019, Radack appeared to relent, tweeting the following:

> Since April 2018, I have been involved in litigation with Trevor Fitzgibbon. We have amicably resolved our differences. As part of the settlement, I retract and withdraw every allegation and statement I have ever made against Trevor Fitzgibbon.

Still, the damage had been done. There's no way to go back. FitzGibbon lost his business, split from his wife, and was too discredited to find work. If prospective employers or clients were to conduct an online search using his name today, it is unlikely they would see Radack's retraction or stories clearing his name. But the old stories with the unfounded accusations are still there. (Radack declined my interview requests.)

"I couldn't defend myself in the press," FitzGibbon laments today. "I was vilified in the national media and on social media. And the accusers, and whatever political machine came after me, used it to poison the water to make it almost impossible for me to get work."

That brings us to another recent #MeToo scandal, one that broke at my alma mater: CBS News.

CBS and the Untold #MeToo Story

It is November 2017, and the *Washington Post* has just published an article lodging allegations about lewd behavior by longtime CBS and PBS host Charlie Rose. More details come in a *Post* follow-up in May 2018. Then in July 2018 comes a blockbuster report in *The New Yorker* by former NBC reporter Ronan Farrow. Farrow's story is a mishmash of divergent innuendo, rumor, and allegations against several CBS men ranging from boorish and sexist behavior to mismanagement, cover-ups, and sexual misconduct. And oddly swept up in the resulting news coverage is *60 Minutes* executive producer Jeff Fager.

There is a lot of off-narrative background behind the CBS scandal that you probably never heard—until now. This is the untold story of how some people believe Fager's enemies and competitors managed to weaponize the #MeToo narrative to accomplish his professional destruction. With help from the media, these players exploited the negative attention on CBS as an opportunity to tar and controversialize Fager. How? Complaints about his management style were magically morphed into the notion that he had turned a blind eye to sexual harassment, which magically morphed into the implication that Fager himself had engaged in rampant sexual misconduct. It is the power of The Narrative that made it possible for unsubstantiated and slanted claims against Fager to reach national news status.

Going back to the beginning, Farrow's earlier reporting in *The New Yorker* had just won a Pulitzer Prize for "exposing the decades-long sexual predation of the movie producer Harvey Weinstein." His next big article about sexual escapades at CBS is well researched. In it, he gives voice to a wide range of #MeToo accusers. To sum things up: Six women claimed that CBS CEO Les Moonves had sexually intimidated or attacked them. More than a dozen CBS or ex–CBS employees claimed that various levels of bad behavior had occurred at CBS News and *60 Minutes*. And they claimed Fager, as the head of *60 Minutes*, tolerated harassment. The article also includes other vague implications about Fager's own personal conduct. An anonymous former CBS employee complained about Fager's "behavior at parties." Several anonymous ex–CBS employees claimed Fager touched employees in "a way that made them uncomfortable." It is left to readers' imagination as to what that could possibly mean. The vagueness and anonymity of the accounts mean there is no meaningful way for Fager to refute them.

"I'm a little surprised [Fager] got lumped into all the other stuff," a longtime CBS News female producer tells me. "Before the Ronan article, nobody at CBS was saying anything about [Fager] regarding sexual harassment."

To me, it seems like a giant stretch for Farrow to group impre-cise innuendo and management complaints about Fager in with al-legations about Moonves, who was accused of being a serial sexual predator. The Fager part of Farrow's #MeToo story is bereft of the normal elements required in responsible journalism. In his story, the accusers are anonymous. The accusations are so nonspecific that it isn't clear that, even if they were true, Fager had done anything improper, let alone illegal. Taken in isolation, the claims about Fager would not have merited national news—prior to #MeToo. Yet there they are, all wrapped up into one big, nasty #MeToo ball.

And it is about to get worse.

About two months later, on September 9, 2018, *The New Yorker* and Farrow publish a follow-up story. That one compounds the charges against Moonves with allegations of specific sexual assaults. (Shortly after the article was published, CBS announced that Moonves had left the company.) Once again, the article throws in a mention of Fager. Again, it feels incongruous. *One of these things is not like the others*, I think to myself as I read the new account. This time, a former intern claimed that some years earlier, Fager "grabbed her ass" at a work party. Paradoxically, considering the context, the woman added that she "didn't think Fager was propositioning her." She says she took the alleged gesture as a *"Welcome to 60 Minutes, you're one of the gang now."* Amplifying the confusion over what we were supposed to make of all of that, the woman went on to tell Farrow that sometime after the alleged groping, all had apparently been well—because she and a fellow intern later invited Fager to lunch and were "excited" when he accepted. Again, I knew that in isolation, these allegations about Fager would not warrant a national news story. But when jumbled together with charges that he had cultivated a "frat house" environ-ment at *60 Minutes* and tolerated harassment, and when recounted adjacent to the horrible tales about Rose and Moonves, it folded into a narrative that was deemed fit to print.

"Jeff could be harsh at times as a boss. But it was equal opportunity whether you were a man or a woman," says a woman at *60 Minutes* who tells me she was surprised to see Fager tarred by the #MeToo narrative. "The amount of women Jeff allowed to split their job after they had babies—Jeff did a lot for women. He promoted a lot of women; he hired a lot of women."

Says another *60 Minutes* woman, "I know there were people who stood up for Jeff in interviews [with Farrow] against charges of sexual predation. But the stories just got published despite that. I think it was grossly unfair."

Here we see that when the narrative train begins chugging down the news track, it can be impossible to alter its path. As other reporters copy, quote, repeat, and amplify the former intern's claim about Fager in *The New Yorker*, the tale miraculously grows taller on down the line. Now a single, unverified allegation gets reported as if it were only one among many. A *New York Post* headline declares that the intern is "another" woman accusing Fager of sexual misconduct yet there had not been a previous accusation. Mediaite chooses a similar headline, referring to "New Sexual Misconduct Allegations" against Fager, incorrectly stating he had previously been "accused of groping co-workers"—plural. But he hadn't.

Fager was soon fired from CBS—but not for the alleged mismanagement or misconduct. It was over a sharp text message he sent the CBS reporter assigned to cover the scandal at her own company.

"If you repeat these false accusations without any of your own reporting to back them up, you will be responsible for harming me," Fager threatened in the text message. "Be careful. There are people who lost their jobs trying to harm me, and if you pass on these damaging claims without your own reporting to back them up, that will become a serious problem." The text sealed his fate. CBS president David Rhodes called Fager in and told him he was being fired for a text that violated company policy. In the final analysis, the innuendo itself was not enough to merit Fager's termination, but his attempt to defend himself proved to be. One could say that unsubstantiated and exaggerated claims, coupled with sloppy reporting, drove Fager to a breaking point where he sent a text message that cost him his job.

Fager issued a statement after he was fired, saying his text message was simply a demand that "she [the CBS reporter] be fair in covering the story. My language was harsh and, despite the fact that journalists receive harsh demands for fairness all the time, CBS did not like it. One such note should not result in termination after 36 years, but it did."

There is more behind-the-scenes context to the story of how Fager was taken out. There had been a long-standing rivalry between Rhodes and Fager. Although none of my sources cited any direct evidence, some CBS insiders say they suspect Rhodes helped unleash the #MeToo fury at the network, in part to take Fager down.

For his part, Fager suggested that the claims against him were orchestrated by a group of people who saw an opportunity to settle old scores over getting fired or passed over. And there is another interesting twist. More than one CBS insider told me they wondered if Farrow had set Fager in his sights because Fager refused to interview Farrow for a *60 Minutes* position several months before.

"David Rhodes wanted Jeff [Fager] to meet with Ronan Farrow [then at NBC] about a job at *60 Minutes*," confirms one CBS official. "Jeff didn't think Farrow was ready and didn't want to dangle the job at people who had no chance of getting on at *60 Minutes*. So he didn't meet with him."

Another CBS insider who worked with Fager expressed the opinion that outside reporters covering the CBS scandal railroaded Fager to help fulfill The Narrative that "powerful men at CBS were all in it together."

"In fact, Jeff doubled and tripled the amount of women, promoted them to senior positions across the network," says that insider. "But all that takes away from the narrative that CBS was a bad place for women. If you don't weigh in with the narrative, then you're portrayed as 'blind' to the problem. They can turn any fact into part of the narrative without actually realizing they just got it wrong. It's an echo chamber, and they just repeat without verification."

Whatever the case, news of Fager's firing from CBS News set off a new media firestorm. Some reporters took his termination as the green light to push the bounds of The Narrative further, even if it cast aside normal journalistic standards. You could say that the news coverage really went off the rails.

Associated Press (AP) joins other media that incorrectly morphed the allegation of Fager making an "ass grab" into multiple "reports that he groped women at parties." AP also states that Fager had been fired in the wake of a "sex abuse probe," which implies, to those who don't know better, that Fager himself had committed sex abuse. Before long, Fager's name and face are splashed on lists of famous alleged predators accused of rape and other criminal acts, such as Weinstein and Kevin Spacey.

A woman who is still at *60 Minutes* tells me, "I thought that the suggestion that Jeff was a serial predator was outrageous . . . unfair and, in my view, untrue." But that side of the story, said several CBS people, isn't what Farrow wanted to hear.

Various CBS insiders provided me with their own theories as to how and why they think the #MeToo narrative was weaponized against Fager. "People were working every day to bring him down," says a familiar on-air face at CBS News. "There was a great deal of competition and power struggles. The #MeToo story was just a way to seal the deal against Jeff."

When I first reached out to Fager after his departure from CBS, he told me he didn't see the point of trying to mount a detailed public defense. He viewed it as a no-win situation. If he defended himself, some would surely misconstrue it as his playing the victim or being unsympathetic to the #MeToo plight, he said. The Narrative was simply too formidable to take on. That's the inherent problem with any narrative: it carries such momentum and force that denying it feels futile. It only fuels The Narrative further. Plus, Fager was holding out hope that an independent investigation CBS had commissioned would clear him of sexual misconduct as well as the charges that he had tolerated harassment.

Four months later, the CBS-commissioned independent probe was finished, but the final report would never be released. Those mentioned in it were not given a copy or the chance to respond to its conclusions. But somebody did leak select findings to the *New York Times*. And they proved to be far from what the media had suggested about Fager. The findings included:

- The media exaggerated some claims against *60 Minutes* employees, including Fager.
- The misconduct was "not as severe as the media accounts or as severe as the sexual misconduct that occurred during the Don Hewitt era at *60 Minutes*."
- There was no evidence Fager had been "aware of the severity of Mr. Rose's inappropriate conduct."
- Fager "demonstrated sensitivity and support for working women."
- Under Fager's leadership, *60 Minutes* "promoted more women to producer and to other senior roles."

That is a far cry from the initial, near-hysterical accusations published by the media. So the question to be asked in the age of The Narrative is: *If you heard so much about the original, lurid claims, why did you hear so little about the counterpoints and the results months later?*

"That's the power of the narrative," says a female Fager supporter at CBS. "It swoops up anything in its path. You want to set the record straight, but everything gets turned around on you. You are told you're part of the problem. It's hard to escape it once it's out there, especially with this particular movement. Traditional reporting goes out the window. They don't care to make sure that their source doesn't have some kind of hidden agenda."

"The Narrative was #MeToo. And it was weaponized [against Fager] by people who had axes to grind," says a CBS insider.

CHAPTER 4

When Narratives Collide

On Thursday evening, March 5, 2020, there is a sad, quintessential case of self-inflicted, lost media credibility. MSNBC news anchor Brian Williams and *New York Times* editorial board member Mara Gay are commenting on money in politics. Specifically, they are referring to a tweet they read on Twitter. (First mistake.) The tweet claims that former Democratic candidate for president Michael Bloomberg could have made everyone in the United States a millionaire with the money he'd spent on political ads in 2020: $500 million.

"Somebody tweeted recently that actually with the money he spent, he could have given every American a million dollars," declares Gay on Williams's program *The 11th Hour*, flashing a big grin.

"I've got it, let's put it up on the screen," says Williams excitedly. "When I read it tonight on social media, it kind of all became clear."

The tweet in question was posted by someone named Mekita Rivas, who describes herself as a journalist with degrees in journalism and English, and bylines in *Glamour* magazine and the *Washington Post*. Williams reads the tweet on the air, tossing in a bit of his own commentary: "'Bloomberg spent $500 million on ads. U.S. population: 327 million.'—Don't tell us if you're ahead of us on the math— 'He could have given each American $1 million and have had money left over.'" The Rivas tweet goes on, "I feel like a $1 million check would be life-changing for most people. Yet he wasted it all on ads and STILL LOST."

Williams concludes, "It's an incredible way of putting it!"

Gay parrots, "It's an incredible way of putting it. It's true. It's disturbing. It does suggest, you know, what we're talking about here, which is there's too much money in politics."

There is one big problem with the Williams-Gay segment. They managed to miss a big, fat miscalculation in the tweet they'd cited; it was off by a factor of about 1 million. In fact, if Bloomberg had spent $500 million, it would have been enough to make *five hundred* Americans millionaires, not 327 million—a difference of six zeroes.

What somehow escaped Williams and Gay was not lost on the viewing audience. The mistake was widely lampooned, and the on-air personalities later apologized.

"Buying a calculator," Gay tweets after the embarrassment.

"Please buy two," replies Williams.

It would be easy to write the whole thing off as a silly gaffe. But I think this says a lot about what has happened to the media and why people do not trust the news. There are so many issues raised:

- A "journalist" issued the initial tweet making the false claim.
- Other journalists accepted the tweet at face value without bothering to do the simplest fact-check.
- They presented the information on a national news program without verifying it, even though verification would have taken nothing more than a moment of reflection.
- A shocking number of hands touched and advanced the mistake without catching it. From MSNBC staff who produced the Williams-Gay segment to graphics personnel who built the visual representation of the tweet, nobody even did the third-grade math to see that it was untrue.

The biggest problem of all is that these are journalists who ask the public to believe, on a daily basis, that *they* can be trusted to present accurate information in the news; information that we should be able to believe has been checked and verified. These are reporters who claim to have the market cornered on the truth of a scientific issue or factual controversy. It is reasonable to conclude that they blindly accepted and forwarded faulty material for no other reason than it fulfilled a slanted narrative they wished to believe.

There is growing evidence that on a daily basis, we in the news media utterly fail to see the role we play in undercutting our own credibility. This is further demonstrated by an insider's view of four recent Washington, DC, journalism functions I attended. Come with me.

The first I want to discuss occurs on March 13, 2019. The event is the Radio Television Digital News Foundation (RTDNF) dinner. A couple of weeks before, I had been at the famous Gridiron Club Dinner, where presidential candidate Amy Klobuchar, a Democrat, was a featured speaker. The Gridiron Club is an invitation-only journalism group with fewer than a hundred members. (I'm not a member, just an invited guest.)

I am surprised to see Klobuchar back now at RTDNF, invited to give a featured talk at a second big journalism dinner in a row. The journalists are honoring her with their First Amendment Defender Award. I wonder how it is that among all the many declared Democrats running for president, Klobuchar has managed to score invites to speak at two coveted DC journalism gigs back to back. Such things don't happen by accident. Somebody influential must like her. We in the media have our "establishment," too.

But if Klobuchar is a favorite among some in the press, others within the Democratic Party are gunning for her. She has recently been hit by a big exposé in the news—one planted by a supposed staffer or someone connected to a staffer. Basically, the news narrative was that Klobuchar abuses her staff like no other member of Congress. That she is so evil, when her staffer once forgot to bring utensils for a lunch on the go, she ate her salad *using a comb*. Then, said the media stories, Klobuchar had the gall to order the staffer to wash the comb. This "news" got treated by some in the press as if it were an investigative report comparable to Watergate.

Combgate!

Klobuchar turns out to be a good speaker. She faces the dinner crowd and displays a big grin. With an animated delivery, she proceeds to make fun of the comb incident. As I listen to her self-deprecating take, I'm pushing around a sliced fingerling potato on my plate and thinking about how it is a brave new media world in which anonymous gossip such as the comb incident would be treated as front-page campaign news. And how Klobuchar has to use her time to address such a thing—rather than talk about anything important.

I look around the room and recognize a few faces. Some incredibly accomplished journalists are here. Each year, I judge the Emmy Awards and continue to see evidence of impressive journalism being committed at many national news outlets from ABC and CNN to PBS, Vice, and HBO. But there is a difference today. The competition to conduct groundbreaking investigations on diverse topics is not as fierce as it used to be. High-quality reporting is not as easy for viewers to find as it was a decade ago. Today, unless journalists are reporting negatives about President Trump, their work isn't likely to get front-page attention, be picked up by the rest of the national press, or become super-amplified on social media.

I can't help but think that here we are, closer to the end of President Trump's first term than the beginning, and we in the media have utterly failed to follow through on our promise to self-reflect upon our missteps during Campaign 2016. To accept what we might have done wrong. To reassess some of our practices that took us off track when we began blending opinion and fact. Where we routinely commit the sorts of mistakes that a novice journalism student shouldn't make in college.

The point is driven home as the night drags on. RTDNF begins to recognize journalists it considers to be First Amendment warriors. One after another, a presenter or honoree rises to enthusiastic recognition and applause. We pat ourselves on the back so hard we could risk breaking an arm. We speak to how proud we are that we supposedly just report the facts, not opinions. *We follow the story no matter where it leads*, we tell each other. I don't doubt that's true of some in the room. But I also know it isn't true of all. And that elephant looms large.

Washington journalism dinners have their own controversial past. It hasn't even been a full year since the White House Correspondents' Dinner presented the "comedian" Michelle Wolf as featured entertainment. Wolf managed to stun even the hardest-core liberals among the audience of three thousand with cringeworthy attacks on President Trump, his staff, and even his family—some of whom were in the room. She drew audible gasps when she set her sights on White House press secretary Sarah Sanders. Sanders was seated at the head table, close enough to have flung a forkful of potatoes at Wolf if she'd tried.

"Every time Sarah steps up to the podium, I get excited," quipped Wolf. "I'm not really sure what we're going to get, you know? A press briefing, a bunch of lies, or divided into softball teams. . . . I actually really like Sarah. . . . She burns facts and then she uses that ash to create a perfect smoky eye. Like maybe she's born with it, maybe it's lies. It's probably lies. And I'm never really sure what to call Sarah Huckabee Sanders, you know? Is it Sarah Sanders, is it Sarah Huckabee Sanders, is it Cousin Huckabee, is it Auntie Huckabee Sanders? Like what's Uncle Tom but for white women who disappoint other white women?"

In the fallout after the dinner, former White House press secretary Sean Spicer called Wolf's performance "absolutely disgusting." But others defended and praised Wolf for having pulled no punches about the president and his aides.

Now, a year later, the RTDNF dinner organizers are still mindful of the stinging controversy and trying to avoid a repeat. There is no comedic performance this year. Even so, most of the speakers end up fulfilling the same political function. They make not-so-thinly veiled attacks against President Trump. They are upset that Trump has supposedly labeled us all an "Enemy of the People" and "Fake News." Nobody points out that Trump has repeatedly stated he was referring to what he considers the dishonest press—not everybody. And if we're honest with ourselves, aren't there at least some among us who have proven not to be honest players?

One speaker blames President Trump for the public's low opinion of the news media. But in fact, that train left the station long before Trump. Trump didn't invent mistrust in the media; our own behavior caused it. In 1999, Gallup found trust in mass media at 55 percent. It plummeted to 40 percent in 2014, long before Trump hurled his first famous insults at the news media during the 2016 campaign. He merely jumped aboard the train and made his way to the conductor's car.

I'm thinking about this as the second-to-last speaker at the dinner nears the end of his speech. His face turns dark. He's clearly rehearsed his part, perhaps while looking into the mirror. He knows it will play well with this audience.

"I am not fake news," he booms in a carefully intoned baritone. "I am not the Enemy of the People." He continues on, criticizing President Trump for being uniquely dangerous to our Democracy, our republic, free speech, and the free world. "We report the Truth," he insists as he frowns and begins breathing heavily. "We are the fact tellers! The truth tellers!"

The dialogue in my mind involuntarily calls up the terrible series of factual errors we've made—not necessarily the people in the room, but some supposedly top reporters in our industry. We *haven't* always told the truth. We *haven't* always gotten it right. Too often, we aren't presenting just the facts. Still, the crowd pays rapt attention to the angry and excited speaker. He concludes with a crescendo and a flourish, and the audience rewards him with a rousing standing ovation.

Yes, the dinner is a night in which we, the news media, celebrate ourselves. But it's hard for me to swallow the fact that we seem incapable of even a scintilla of introspection.

So what should have been said at the dinner? Maybe something like "It's a time to celebrate and defend what we do when we are at our best. But we can't do that without noting that we have not always been at our best. Some of us have suspended our normal ethics and practices. We've blended opinion and reporting. We've self-censored people and topics. We've stepped in to try to shape public opinion rather than report the facts. It is only with this recognition of the fact that we have a problem that well-intended, serious journalists can begin to solve it."

As I finish off my chocolate dessert, I wonder if I am the only one at the dinner who has that running dialogue in my head.

Seven months later, it's October 24, 2019. There's another dinner sponsored by the Radio and Television Correspondents' Association at The Anthem, a magnificent waterfront auditorium and music venue in Washington, DC. The journalist and author Jon Meacham is giving the keynote speech. It becomes another uncomfortable and telling moment.

Don't misunderstand me; Meacham is a good speaker. And I think the audience enjoyed listening to the first part of his talk. It was sprinkled with presidential history and humor. But the night takes a different turn in the latter part of his presentation. He includes President Trump in the category of what he describes as "lawless presidents." Then he advocates for Trump's impeachment. More than that, he issues a call to action, telling his audience of journalists that we will be judged for whether or not we stand up now to save our republic from Trump and all the horrors he's wrought. "*Stopping Trump—it's up to all of us,*" he lectures the roomful of reporters.

Meacham is perfectly entitled to express his anti-Trump opinions. And I might have appreciated listening to them under different circumstances. But he should not have been invited to present his incendiary and slanted political views to a group of journalists who are already facing public criticism and scrutiny for their political bias.

I sense that I'm not the only one in the room who is uncomfortable with this choice of content. As he goes on, I see people sharing glances and heads leaning toward one another to exchange whispers at nearby tables. After Meacham concludes, I visit the ladies' room and hear two young journalists I'd never met expressing their dismay. They steal a glance at me. One of them shakes her head and whispers, "Inappropriate." We briefly chat and agree that the same folks who consider Meacham's speech perfectly reasonable fare for one of the year's top news journalism dinners would never dream of inviting a speaker who is the equivalent of someone just as biased on the other side. We know that the mere suggestion of an invitation to someone like that—someone who doesn't hate President Trump— would prompt threats by some of our fellow journalists to boycott the dinner.

This example drives home how pervasive certain biases are in some corners of the news media—so much so that the bias is considered the default position. When it comes to news reporting, the center has been dragged so far left that a neutral posture is now viewed as right wing. Liberal or anti-Trump views—those are considered good, truth-telling journalism. At least that's what the afflicted seem to believe. But raise questions about fairness or consider alternate viewpoints—that simply proves you're the one who's biased. Maybe even (gasp!) conservative. (Although you're not.)

This syndrome builds upon itself with assistance from those working so hard to whittle down the universe of information and views we see on the news and online. As conservative voices are purged, the realm of acceptable views is jolted to the left. Moderates who are close to the center begin to look conservative, at least in relative terms. A number of "traditional" liberals have written about this phenomenon. They wonder what happened to a time when journalists, above all, fiercely encouraged a wide range of views and rejected the notion of filtering or censoring information in order to tell the public what to think.

It reminds me of an acquaintance who, not long ago, was in

charge of booking the featured host for yet another Washington, DC, journalism-related dinner. Typically, as with most of these affairs, the featured speakers, comedic entertainment, and honored guests are liberal-leaning. But this particular year, my acquaintance scores a celebrity to serve as emcee: the actress Janine Turner, formerly of the hit CBS television program *Northern Exposure*. My acquaintance is a fan of hers and is delighted that she has agreed to the booking. But then the story turns sinister. You see, someone learns that Turner is a devout Christian with mildly right-of-center political views. Before long, the organizers of the event make a strange and sudden decision. They decide that the dinner will no longer require the services of an emcee and host. So naturally, Turner's services will no longer be needed. She's canceled.

When we cannot see or admit the bias among ourselves, it is little surprise that it shows in our reporting, too.

The last journalism dinner I will mention here is the annual RTDNF dinner on March 5, 2020. It is now two full years after the Michelle Wolf disaster, and I notice that the event appears much smaller in terms of attendance. There is another liberal Democrat, Senator Richard Blumenthal, following in Senator Amy Klobuchar's footsteps to receive the group's First Amendment Defender Award. Blumenthal gives a brief speech declaring that the First Amendment has never been more threatened than it is today. I agree, but for different reasons. Blumenthal goes on to comment that he finds it outrageous that he had to propose passing a Reporter Protection Act in Congress in the face of dangers posed by a "public official" (i.e., Trump) threatening the press and bullying reporters.

As I listen in the audience, a first-time attendee at my table leans over and asks me in a whisper whether the organizers have ever bestowed the First Amendment Defender Award to someone on the other side of the political aisle. She seems surprised at the obvious slant. She must not be from around here. I don't even have to look at the program's history to know the answer. These events typically lean left with an occasional token conservative or liberal Republican thrown into the mix. A quick check shows that my guess is correct: the only three political figures honored with an RTDNF First Amendment Defender Award in the past three years are all liberal Democrats. One might deduce that the organization finds no value in presenting the appearance of bipartisanship. Or that it has never occurred to them to do so. Or that they believe there simply are no Republicans worthy of recognition as strong defenders of the First Amendment. The first-time diner's question about the slant of the dinner transports me back in time. I, too, was once surprised at blatant partisanship demonstrated and embraced by national reporting organizations. But that was a long, long time ago. I've grown accustomed to the notion that in this crowd, a liberal tilt is actually considered to be the center. The blatant partisanship is viewed as nothing to debate or self-reflect about; it is simply seen as the undeniably correct view.

As the dinner draws to an end, the closing speech is given by an honoree who is a former news executive. He continues the night's theme by declaring "We cannot find inspiration in our leadership [i.e., Trump]." So he advises us to look for inspiration not from anyone born in America but from those at an immigration and naturalization ceremony. That is where, he declares, we will find "all good people." As this now retired big-city news executive speaks of his new life out in middle America, I almost think we are about to hear a teeny hint of introspection that is too often lacking in journalism. He tells the audience of news reporters, producers, and news executives that we would not believe how many people in middle America are critical of what we, as reporters, do! But instead of concluding that we might want to look at where we might be falling short, he advises a different tack: he says we must work harder to convince the public that we are working in their best interests. *"They just do not understand,"* he tells us. Among his advice, he says we should create daily promotional campaigns that explain to the public where they are wrong about us.

Collision Course

Now you understand a little more about the starting point for many journalists. This is part of what makes our reporting so susceptible to the influence of special interests. It is why, in this realm, slanted can seem perfectly straight.

It is difficult to overstate how widespread the manipulation of news is, based on what does or does not fit a particular narrative. A big way we in the media sabotage our own credibility is by adopting the language of propagandists in our reporting.

One expects propagandists to devise and deploy their own phrases to influence public opinion. But we in the press are supposed to resist populating our news reporting with charged and pejorative terms. Stated differently, commentators and operatives seek to incorporate their invented language into our daily lexicon. But news reporters should not help them do it. We should stick to using factual descriptions or at least attribute the propaganda terms to their sources rather than adopt them as if they are our own.

It is not only a good practice; it keeps our reporting more accurate than it would otherwise be. In some instances, news reporters now commonly use phrases that are factually incorrect when examined using a neutral reporter's eye. Here are seven examples of propaganda terms that have successfully wormed their way into the news.

1. **"-PHOBIC" AS IN "HOMOPHOBIC, ISLAMOPHOBIC, TRANSPHOBIC":** *Phobic* is defined as "having extreme fear or aversion." But today, some news reporters commonly use the term to describe people who differ with others on certain policies or who may dislike given behaviors—even though they are not "afraid" or "fearful" of them. One might correctly be able to report that people or views are intolerant or even hateful, but from an accuracy standpoint—which matters to our credibility—it is not correct to characterize them as phobic. Example: asking if someone

was "born a boy or a girl" is deemed "transphobic" by the website Mashable, even though the question itself does not connote fear or hate. The same is true about the phrase "he or she," which some activists incorrectly claim is transphobic.

2. "DEBUNKED": This word was rarely used in news reporting until a few years ago. That's when propagandists began deploying the term to discredit theories, stories, and science with which they disagree. In fact, when special interests launch this word, it often means the opposite is the case: the targeted idea has not been debunked at all. Oftentimes, the idea in question is a subject of legitimate dispute or has actually proven to be true. Therefore, it is often inaccurate for news reporters to jump on the "debunked" bandwagon. The term "bogus" also falls under this category. Example: On February 17, 2020, Paulina Firozi of the *Washington Post* falsely declared the idea of the virus coming from a Wuhan laboratory to have been "debunked." It had not been debunked. In fact, an April 14, 2020, article in the *Washington Post* debunked the earlier *Post* article's claim that the Wuhan tie had been debunked, by acknowledging that the idea was still under wide consideration. "[M]any national security officials have long suspected either the WIV or the Wuhan Center for Disease Control and Prevention lab was the source of the novel coronavirus outbreak," reported the *Post*'s Josh Rogin.

3. "FAKE NEWS": This term was popularized in its modern context by liberal interests and news reporters in 2016 in order to steer the public away from certain ideas and information. But after it was co-opted by Trump and his supporters, news reporters declared that it was time to "retire" the term. They accomplish the same propaganda goal through "fact-checks" that label information as "false," even when it may not be. For example, President Trump opposed mail-in ballots for the 2020 race, citing the opportunity for widespread fraud. A chorus of news reports declared that to be factually incorrect and said the president's

concerns were "without reason," though that is only their personal opinion and they cannot foretell the future.

4. "ANTI-IMMIGRANT": Advocates who support illegal immigration have successfully supplanted the phrase "illegal immigrant" with "migrant" or "immigrant." They have likewise replaced "anti–illegal immigrant" with "anti-immigrant," as if the terms are equivalent. In fact, from a factual standpoint, they have two entirely different meanings. Example: Julissa Arce in *Time* falsely quoted Trump from a May 2019 rally as if he had spoken of "migrants" in asking "How do you stop these people? You can't." But Trump was specifically referring to immigrants attempting to illegally cross the border from Mexico into the United States, not legal migrants or migrants in general. News reporters should not fall into the trap of using "anti-immigrant" to describe those who favor immigration but are against illegal immigration.

5. "ANTI-SCIENCE": Propagandists working for a variety of special interests have codified this term for use against those who question their claims, theories, and findings. It is frequently a misnomer because the questioners are rarely "against science," as the term states. They simply differ on which science is accurate, or they have different interpretations and conclusions. Example: On May 24, 2020, the *New York Times'* Knvul Sheikh declared that the debate over wearing masks to prevent coronavirus had been "settled" (she says masks work)—less than three months after she coauthored an article quoting public health officials who said they were not effective. Meantime, many authorities, including a World Health Organization official, said the jury was still out on masks. The same goes for the phrase "anti-vaccine," which reporters routinely use incorrectly to controversialize or disparage scientists and others who investigate vaccine safety issues. Much the same can be said about "settled science." Science is rarely settled, and legitimate scientists or reporters rarely claim that it is. It is a term of propagandists.

6. **"DENIER"**: Whether it is "climate denier" or "science denier," some reporters misuse the pejorative term to describe people who differ on scientific theories, proof, and conclusions but do not deny that a climate or science exists. Example: In October 2019, a group of five hundred international scientists wrote the United Nations, saying that there was "no climate emergency," urging "a climate policy based on sound science," and asking for more voices to be heard in the debate. The website Quartz characterized that as "climate change denial" and went so far as to claim that "most of us are at least one type" of "climate change denier." I, for one, do not deny that the climate is changing, and I do not believe I know anyone who does.

7. **"ANTI-GUN"**: Not all policies restricting gun use can accurately be defined as anti-gun. In fact, many gun control advocates also support gun rights; they just differ on the extent of the rights.

Here is some insight: when you hear reporters use propaganda words and phrases like these, it often signals that you should take a second look at the information they are presenting and ask yourself—*Who might be pushing a narrative?*

Now you have a basic primer on some of the prevalent thinking and flaws within my industry. It helps explain the mentality of some news insiders. They are so influenced by ideology and conflicts of interest that they do not seem to notice they are reporting in ways that don't make logical sense. They are so ruled by propaganda over facts and reason that they take positions that are self-contradictory.

So what happens when narratives collide? Chaos and uncertainty reign.

That was the case when the #MeToo narrative collided with powerful interests that apparently preferred to tamp down the exploits of a well-connected sex offender.

This example is brought to us with help from Project Veritas, a conservative-leaning nonprofit dedicated to investigating corruption, dishonesty, waste, and fraud. The group, founded by James

O'Keefe, is both lauded and criticized—depending on where one sits—for its undercover videos exposing left-wing bias in the media and scandals such as Planned Parenthood officials discussing the sale of aborted fetus parts.

On November 5, 2019, Project Veritas publishes a shocking videotape of an ABC News reporter named Amy Robach. In the video, Robach claims her own network had engaged in a sort of cover-up of crimes committed by the sex offender Jeffrey Epstein. Epstein was the sex trafficker who met with an untimely death in a New York prison earlier in the year while awaiting trial on new charges. Prior to his death, which was officially ruled a suicide, he was said to have had the goods on many important and powerful world figures. Polls show that most Americans believe Epstein was murdered.

Anyhow, in the video slipped to Project Veritas, Robach is seen and heard on camera chatting with someone off-screen in between her televised news appearances. She is complaining that, in 2015, she taped a blockbuster exclusive interview with an Epstein associate named Virginia Roberts Giuffre but ABC blocked the story from airing.

The scandal harkens back to my time at CBS News when a number of us noticed some managers making news decisions we felt were contrary to good journalism as well as the public interest. The syndrome escalated during my last few years at the network, ultimately leading me to ask to depart in 2014 ahead of my contract expiring.

Sometimes, I believed, the conflicted news managers were influenced by their own ideologies and worldview. Other times, I learned that powerful corporate or political figures had reached out to important people at CBS to push narratives and block any contrary information. Some managers even told me they worried that if they were to air one of my hard-hitting news stories on a particular person or topic, it would hurt our chances of getting fluff celebrity and political interviews in the future. A CBS News executive once confided in me that Obama administration officials had threatened to pull CBS out of the "rotation" for their next "handout" interview if *CBS*

Morning News dared to air one of my investigative reports. A handout interview is an interview the administration arranges and offers to the network news outlets. (If you were under the impression that the interviews we conduct are our own ideas based on news value, this might surprise you. Too often, it is not the case. But I digress.)

The dynamic whereby newsmakers call the terms of our coverage by threatening to withhold access to a handout interview never made sense to me. Their arranged interviews and anonymously leaked "scoops" often amount to slanted, unverifiable information that they want publicized to shape public opinion. It is much closer to propaganda than to legitimate news. But for some reason, the suggestion that we might lose our chance at a handout interview, or that our "access" to administration figures could be jeopardized, leaves us shaking in our shoes. *We mustn't risk being skipped over for the next fluff or celebrity interview!*

This is a problem not only at CBS. I attended a big national journalism conference shortly after my departure from the network in 2014. While there, I spoke one-on-one with high-ranking executives from CNN, ABC, and NBC. Each confessed to me that his network, too, had buckled under similar, explicit threats from Obama officials, and Bush officials before that. They don't like it any more than I do. But somehow they felt powerless to stop it. We allow ourselves to be manipulated.

There are countless other ways reporters and news outlets are pressured to block stories that are contrary to the narratives of powerful interests. In 2011, I was headed on a work trip from Reagan National Airport just outside Washington, DC, when I got a call from one of Katie Couric's assistants. At the time, Katie was the anchor of *CBS Evening News.*

"Remember you did that story on the Copenhagen climate summit a year ago?" Katie's assistant asked.

I remembered the story well. It was one of my classic "Follow the Money" series, ever popular with viewers. This one had aired on January 10, 2010, and was titled "Copenhagen Summit Turned Climate

Junket." I followed it up with a sequel a few weeks later called "Congress Went to Denmark, You Got the Bill." The two stories exposed a large congressional delegation—both Democrats and Republicans—spending a lot of taxpayer money traveling to a climate summit in Denmark even though they knew in advance that no major deal would be signed. The details of who attended and how much it had cost were supposed to be public information, but the official disclosures had not yet been filed when I sought them. And congressional leaders would not release them to me when I asked. So I did a little detective work and pieced together data by developing sources and gathering documents from individual offices of members of Congress. The point of the reporting was to highlight the inherent hypocrisy revealed by environmentalists burning so much oil to attend a global warming gathering where nothing significant would be determined, and also to uncover how much tax money had been spent to boot.

Quoting from my reports:

[Our] cameras spotted House Speaker Nancy Pelosi at the summit. . . . House Majority Leader Steny Hoyer and embattled Chairman of the Tax Committee Charles Rangel were also there. They were joined by 18 colleagues: Democrats Waxman, Miller, Markey, Gordon, Levin, Blumenauer, DeGette, Inslee, Ryan, Butterfield, Cleaver, Giffords; and Republicans Barton, Upton, Moore Capito, Sullivan, Blackburn, and Sensenbrenner . . . the congressional delegation was so large, it needed three military jets—two 737's and a Gulfstream Five. Up to 64 passengers traveling in luxurious comfort. Along with those who flew commercial, we counted at least 101 Congress-related attendees. . . . As a perk, some took spouses. . . . Rep. Gabrielle Giffords was there with her husband. Rep. Shelley Moore Capito was also there with her husband. Rep. Ed Markey took his wife, as did Rep. Jim Sensenbrenner. Congressman Barton, a climate change skeptic, even brought along his daughter. . . . Three military jets at $9,900 per hour: $168,000 just in flight time. Dozens flew commercial at up to $2,000 each. 321 hotel nights

booked, the bulk at Copenhagen's five-star Marriott. Meals added tens
of thousands more. . . . They produced enough climate-stunting carbon
dioxide to fill 10,000 Olympic swimming pools. Which means even if
Congress didn't get a global agreement—they left an indelible footprint
all the same.

Fast-forward to about a year after those CBS reports. Tragedy struck
one of the Copenhagen climate summit attendees, Congresswoman
Gabby Giffords. On January 8, 2011, a maniac shot her and eighteen
others in Tucson, Arizona. Six people were killed, and Giffords was
critically wounded. That dreadful event figures into the phone call
I got from Katie's assistant a few months later at Reagan National
Airport.

"I need you to justify why you mentioned Giffords in your Copenhagen climate summit story," Katie's assistant tells me on the phone.

Justify it? My mind races to digest what she's asking and discern
where the conversation might be going. I tell her something along
the lines of "It was a 'Follow the Money' story on all the tax money and
carbon Congress expended by taking so many people to an environmental summit where no major agreement was going to be signed."

"But *why* did you mention *Giffords* in the report?" she persists.

"Because she was one of the twenty-one members who attended. I
don't understand the question."

"Well, you see," the assistant explained, "Katie's trying to get the
first interview with Gabby after the shooting, and her staff is still
really mad about that story you did. Apparently, it really impacted
her last election. They're very upset."

"And?" I reply.

"And it's making it hard for us to get the interview."

I hear my flight called for boarding, and I still am not sure why I
feel as though my arm is getting twisted over a thoroughly reported
story that aired a year before.

"If it helps," I suggest, "just tell them, 'That was Sharyl's story'
and that 'Katie doesn't control what Sharyl does.'"

"It's not that easy," says the assistant. "After all, it's Katie's name on the program."

I finish the call and share what's been said with my producer. Why should reporters in the field have to wonder if their legitimate news story might be second-guessed later as part of negotiations for an interview? Such a consideration has a chilling effect on news gathering. My own view is that if we lose an interview because of honest reporting we have done, so be it. If all of the networks simply refused to capitulate to such pressure, newsmakers would not be able to influence the news by threatening loss of access.

One footnote. I don't think Giffords was personally behind the pressure I got in the phone call from Katie's assistant. I later bumped into Giffords and her husband at a dinner in her home state of Arizona, held in honor of a murdered Border Patrol agent, Brian Terry. The Giffordses introduced themselves to me. We shook hands. They were gracious and complimentary. My guess is that a Giffords staffer had been freelancing when previously raising a fuss about my story. I suspect that Giffords's office never really considered granting Katie the first post-shooting interview with Giffords to begin with. Nonetheless, a message had been delivered to Katie that I was a problem. A message had been delivered to me that my good reporting bore negative consequences.

So how does this anecdote jibe with the account of ABC deep-sixing the Epstein story in 2015? Well, in the ABC video leaked to Project Veritas, Robach said higher-ups told her they were worried about losing access if they aired her report. Specifically, she said, they were concerned they would be cut off from future interviews with the British royals (who had reportedly threatened as much). That's because Prince Andrew was allegedly linked to Epstein and some of his crimes.

"[T]he palace found out we had . . . allegations about Prince Andrew and threatened us in a million different ways," Robach is heard saying in the leaked video. "[Virginia] told me everything," she added, referring to Epstein's alleged victim Virginia Giuffre. "She had pictures. She had everything. She was in hiding for twelve years, we convinced her to come out. We convinced her to talk to us. It was unbelievable what we had. [Bill] Clinton. We had everything. I tried for three years to get it on to no avail. . . . I've had this interview for years. . . . And now it's all coming out, and it's like these new revelations, and I freaking had all of it. I'm so pissed right now, like every day I get more and more pissed 'cause I'm just like, oh my God, what we had was unreal."

After Project Veritas posted the video of Robach online, ABC issued a statement denying that the story had been improperly blocked, saying "At the time, not all of our reporting met our standards to air." For her part, Robach issued a separate statement of agreement, saying "I was upset that an important interview I had conducted with Virginia Roberts [Giuffre] didn't air because we could not obtain sufficient corroborating evidence." We are to believe that there was not enough corroboration of Epstein's well-documented transgressions, but the same network required no corroboration at all when reporting salacious allegations against Supreme Court nominee Brett Kavanaugh in 2018. *When narratives collide.*

Make no mistake, this sort of drama plays out with frightening regularity in national newsrooms across the country. You just don't usually hear about it. It is the same pathology I first described in 2014. Gossip blogger Erik Wemple of the *Washington Post* quoted from my writings in an opinion piece about the ABC-Robach controversy. He challenged Robach's publicly issued statement in which she seemed to agree that her story about Epstein had lacked sufficient corroboration in 2011.

"Which Amy Robach do you believe:" wrote Wemple, "The one chatting candidly in her studio, believing that she's just exchanging gossip with colleagues? Or the one who comes to you through a prepared statement distributed by ABC News? In her 2014 book 'Stonewalled,' former CBS News correspondent Sharyl Attkisson inveighed against the mentality that might account for both NBC's and ABC's whiffs [on not reporting about Epstein sooner]: 'Many story topics are selected by managers who are producing out of fear and trying to play it safe,' she wrote. 'Playing it safe means airing stories that certain other trusted media have reported first, so there's no perceived "risk" to us if we report them, too.'"

The ABC scandal demonstrates the sort of tumult that can happen when narratives collide in a big way. National news outlets have proven to be eager, if not delighted, to rush to air with unsubstantiated claims that fit the narrative of powerful men sexually mistreating women. They have jumped on board with the notion that women who make sexual abuse claims "are to be believed" without question. On the other hand, here was ABC in the uncomfortable position of defending why it did not run Robach's story about Epstein and one of his victims. To justify the inexplicable, the network relied on a competing narrative, insisting that Robach had not properly substantiated her story. When it came to this particular news story, ABC was insisting now women who claimed to be victims were not to be "automatically believed."

SUBSTITUTION GAME: In September 2018, the *New York Times* printed an anonymous, scathing opinion editorial, supposedly written by a Trump official, detailing the heroic "resistance" against Trump happening inside his own administration. Contrast the treatment that anonymous figure received in the press to the beatdown of Robach's story at ABC. The anti-Trump "whistleblower" went directly to the press, made unsubstantiated allegations, didn't actually blow the whistle on anything, won widespread coverage, and got a book deal out of it—all because it fit The Narrative.

For our next example of what happens when narratives collide, we go off to the Windy City and examine news coverage about gun violence.

The Chicago Gun Narrative

Perhaps we can agree right off the bat that if thirteen people are shot at one event, it qualifies as a mass shooting—except, perhaps, when it happens in Chicago!

In general, the media embrace The Narrative that mass shootings prove the need for stricter gun control laws. And reporters are often quick to blame mass shootings on those who have fought gun restrictions.

But there is a competing, contradictory narrative that complicates the story line. That narrative says when rampant gun violence occurs in Chicago, Illinois, other factors besides a lack of gun control laws must be blamed (because Chicago already has strict gun control laws). To accomplish The Narrative, news terminology must be adjusted accordingly. When mass shootings occur in the poor, crime-riddled, black neighborhoods of Chicago, they often aren't called "mass shootings." They are referred to as "gun violence" and used to advance narratives about poverty, racism, or the police.

Here are a few examples. One news article about Chicago "gun violence" in 2018 declared, "We have to own up to the racism to really solve the problem." Opinion and news pieces in the *Chicago Tribune* and *USA Today* blamed the gun violence on police for not solving enough crimes. "Last year, the Chicago Police Department solved only about 17 percent of the homicides committed," wrote the *Chicago Tribune*. *USA Today* echoed the sentiment, complaining that "Chicago police solved fewer than one in six homicides in the first half of 2018." The *Chicago Tribune* also blamed poverty, writing that as for "what fuels Chicago's violence . . . solving systemic poverty remains daunting." Axios topped them all, blaming the gun violence on every other factor—except gun laws. Axios cited "Racial segregation, wealth inequality, gangs and the inability of law enforcement to solve crimes," saying all of these things "have fueled the gun violence."

An odd outgrowth of these colliding narratives is that the media highlight certain mass shootings while downplaying or largely ignoring other mass shootings that are equally as destructive.

On December 22, 2019, thirteen people were wounded—four seriously—at a memorial gathering in Chicago. Oddly, not many news outlets wrote headlines referring to it as a "mass shooting."

Among the few that did were some local Chicago media, such as the *Chicago Tribune* and ABC7 Chicago. The national news, including NPR, CBS News, the *Washington Post*, and CNN, avoided the phrase. *USA Today* used the word "mass" but called the tragedy "mass violence" rather than a "mass shooting."

The previous August, there was an even more dramatic example. In one weekend alone, there were four mass shootings, two in Chicago. But only the non-Chicago shootings seemed to qualify as such in the national news media. Here's how it went down.

Two Chicago-area mass shootings happened on August 4, 2019. Dozens of victims were wounded three hours apart. Over the same weekend, there were mass shootings in El Paso, Texas, and Dayton, Ohio. Two plus two equals four. But CNN reported only two mass shootings, mentioning El Paso and Dayton but omitting the ones in Chicago. NBC also headlined its article "2 Mass Shootings in Less than a Day," leaving out the Chicago tragedies. Vox and Forbes did the same. At the end of the year, when ABC added up the nation's 2019 mass shootings, it excluded Chicago's two shootings in August.

One local newspaper in Chicago, the *Chicago Tribune*, goes against the grain and seems to pretty much count all the mass shootings that happen close to home. The *Tribune* reported there were five Chicago-area mass shootings in 2019: the two in December, the two in August, and one in February, when a man shot and killed five coworkers at the Henry Pratt Company plant. Five mass shootings in the same troubled city in a single year would normally merit extensive national coverage. But as far as the national press was concerned, Chicago had just one mass shooting in 2019: the Henry Pratt workplace incident. And many in the media faulted President Trump for that one. Former congresswoman Gabby Giffords tweeted:

> It makes me sick to think that our country has a president who still refuses to acknowledge a real crisis when he sees one. Americans should be able to go to work without fear of being shot. This must stop.

I dug into the question of how reporters might justify these seeming contradictions in which mass shootings get counted and which do not. Various news outlets choose to define "mass shootings" differently. Some reserve the term for when a certain number of people are killed, rather than injured. Others take their cue from the FBI, which at one point defined a "mass shooting" as an attack that kills four or more people. In 2013, it broadened the definition from four to three or more being killed. Despite this guidance, it is farcical to suggest that the media should not consider thirteen people shot in Chicago to be a "mass shooting." But when narratives collide, anything becomes possible.

When Narratives Backfire

There are countless examples of narratives effectively distributed to the consuming public through slanted reporting. But many people are getting wise. And they don't like being taken for fools. There is a growing number of thinking people who look at the facts, deploy their common sense, use their instincts, and apply a little logic when they watch the news. Oftentimes, this leads them down an entirely different road than the one The Narrative intended to point them down. In other words, The Narrative backfires.

There's nothing more deliciously destructive to The Narrative than people who have figured it out. It changes them. I have described it as akin to taking off the foggy glasses you've been wearing and wiping them clean. Suddenly things become clear. People become converted and inspired in this way. They may turn into outspoken advocates, motivated by a desire to open the eyes of those around them.

Candace Owens is one such convert. She told me that once she recognized a major false narrative against Donald Trump, it changed her. And with her metamorphosis, she became the target of false narratives—about her.

Owens is African American. Though she says she'd never voted, she always considered herself to be a Democrat "because black people are supposed to be Democrats."

She tells me in an interview that she "woke up" during the 2016 election. "It was largely due to the fact that the media was going around calling Donald Trump a racist," she says. "And they were really over-playing their hand on this one, for me in particular, because I grew up listening to hip-hop music. Everybody loved Donald Trump! Everyone wanted to be like Donald Trump. Beyoncé and Jay Z were sipping poolside at Mar-a-Lago in their songs. And then suddenly he announces his bid for the White House and black America was supposed to suddenly realize that he was a racist? I was a little too smart for that assessment."

That simple, logical observation set Owens off on a mission to research what else the media might be telling her that is untrue. Her findings, she says, rocked her world. She came to believe that the reason so many blacks vote for Democrats without considering another path is because they've bought into one of the most powerful political narratives of our time, aided and abetted by slanted news reporting.

"If you believe that conservatives are racist and the liberals are your saviors, and suddenly you go through this awakening period, there's a lot of shock," she says.

I ask her what she felt the Democratic Party had done wrong when it comes to African Americans.

"Well, I would almost say in terms of what they wanted to do, they've done everything right," she replies. "What they wanted to do was to marry us to their party by giving us a bunch of handouts, making sure that we would never get ahead, and they were able to do that via the welfare system."

Owens's "awakening" inspired her to start a movement she calls "Blexit," meaning the black exit from the Democratic Party. She encourages other African Americans to lift up the dust skirt on the reliable old couch and see what's really underneath. She asks them to rethink The Narrative that blacks must vote Democrat. She urges them to reject The Narrative that if a black person is a conservative, it is only to seek money or fame.

Democrat Joe Biden focused attention on these common narratives in May 2020. In an interview with an African American radio host, the presidential candidate told the audience, "Well, I tell you what, if you have a problem figuring out whether you're for me or Trump, then you ain't black." That set off a firestorm even among fellow Democrats—of all races—for seeming to imply that black Americans cannot have a mind of their own.

Another lifelong Democrat, Brandon Straka, describes an "awakening" similar to that of Owens. In an interview with me, Straka says that in 2016, he voted for Hillary Clinton. "You know, I was one of these people who was absolutely horrified, crying, shattered, and so upset that Donald Trump had been elected president because I voted for Hillary Clinton," he says.

But he goes on to say that as a gay man, he began observing that the anti-Trump messages he was force-fed on the news didn't match up with what he saw with his own eyes or experienced in person. "While the media is telling me that Donald Trump and his supporters are bigots, racists, homophobes, what have you, I'm not really seeing it in reality anymore. And so I wanted to try to understand where the truth was. I started hearing about these very valid, very real reasons why people supported Donald Trump."

Straka ended up believing that the political party he'd once embraced has abandoned traditional liberal values, such as standing up for free speech and being inclusive. He tells me that the Democratic Party has come to represent hate and divisiveness. He ended up so disgusted by the media and disillusioned by the party that he launched a movement he calls "hashtag WalkAway," as in "just walk away from the Democrat Party."

It's hard to say why some people, such as Owens and Straka, identify and resist the influence of what they believe to be The Narrative, while so many others are perfectly happy to consume spoon-fed views.

On the other hand, there are those who argue that Owens and Straka are themselves part of a narrative, whether they realize it or not. Associate professor of sociology at Georgetown University Corey Fields, the author of *Black Elephants in the Room: The Unexpected Politics of African American Republicans*, says the notion that blacks are fleeing the Democratic Party is a narrative being pushed out by conservatives. He says Republicans embrace converts like Owens because it fits *their* narrative.

"It . . . allows certain messages that pretty much operate as ways of chastising black people to be delivered by black voices," he notes in an interview with me. "Like 'Black people, stop complaining about racism and work harder.' That message coming from a white person would be highly critiqued, strongly critiqued as racist and problematic and troublesome. But that message coming from Candace Owens gets to operate in a way where . . . can you call it racist? Because a black person said it?"

This is one indication of how many layers deep a system of narratives can go. Owens says those who villainize her, often using false information, are serving their own narratives. She does not take the attacks quietly. At a congressional hearing in 2019 about white supremacy, an academic from Stanford University named Kathleen Belew joined Owens at the witness table to provide testimony about their knowledge and experiences. Belew brought up a shooting in Christchurch, New Zealand, on March 15, 2019. A terrorist attacked mosques, murdering fifty-one people. Belew pointed out that the gunman had referred to Owens as a major source of inspiration.

But when it was Owens's turn at the microphone, she fired back. She pointed out that the Christchurch shooter had mentioned many other notables in his "manifesto" and it was unfair to claim that those named were racists or somehow brought on the attack. Owens also took on the fact that her co-panelists, all of them white, were testifying that white nationalism is a major threat in America but could not cite statistics to support that contention.

"I also found it quite hilarious that when asked for actual numbers, nobody here could actually provide them, because [white nationalism] is not actually a problem in America or a major problem or a threat that's facing black America," Owens said at the hearing.

Belew responded by attacking Owens for using the word "hilarious," as if Owens had laughed off racism. Belew then claimed Owens had similarly laughed off being named in the terrorist's manifesto.

"You knew exactly what I meant when I said 'hilarious,'" Owens shot back. "And you just tried to do live what the media does all the time to Republicans, to our president, and to conservatives, but you tried to manipulate what I said to fit your narrative. Okay? I was not referring to the subject matter that is 'hilarious,' I said it's hilarious that we're sitting in this room today, and I've got two doctors and a Mrs., and nobody can give us real numbers that we can respond to so we can assess how big of a threat [white supremacy] is. . . . And the audacity of you to bring up the Christchurch shooting manifesto and make it seem as if I laughed at people that were slaughtered by a homicidal megalomaniac is, in my opinion, absolutely despicable."

The whole back-and-forth at the hearing aired live on CSPAN, and a replay of the clip that circulated on social media had more than 4 million views the last time I checked.

Owens and Straka are deeply influential in certain corners of the Internet and social media—but they remain relatively unknown elsewhere. There was a time when the news media would have trained their attention upon them precisely because their personal stories are unusual and interesting. They are charismatic and well spoken. But in today's news environment, because their stories do not fit neatly into the popular narratives, the popular media downplay or ignore them.

CHAPTER 5

The New York Times

ALL THE NARRATIVES FIT TO PRINT

*Racism is in everything. It should be considered in our science re-
porting, in our culture reporting, in our national reporting. . . . It's
less about the individual instances of racism, and sort of how we're
thinking about racism and white supremacy as the foundation of
all of the systems in the country.*

—NEW YORK TIMES STAFFER, AUGUST 2019

The summer of 2019 sees the public unraveling of the *New York Times*,
once perhaps the most respected news organization on the planet. A
series of internal controversies plays out in the news and on social
media. The controversies expose how The Narrative has taken over
at the Gray Lady and—by extension—the news industry.

The chaos starts in June when the newspaper publishes a dubious rape allegation against President Trump. One might have reasonably expected it to prompt a journalistic discussion over whether
such a story warrants national news coverage, especially in a politically charged environment where so many people have gotten caught
making false or unproven claims. The decades-old rape allegations,
made without evidence by Jean Carroll, a woman promoting a new

book, would not have been deemed suitable material to print in most any credible news publication just a few years ago. The caution bar for uncorroborated sexual allegations used to be set at a reasonably high level. Journalists would not have considered it appropriate to publish serious allegations purely because they would like to believe an accuser or wish for the claims to be true.

But today, things are different. The media have unsheathed all their daggers to destroy Donald Trump. In this environment, unsubstantiated rape claims merit a write-up in one of the world's most important print news publications. It fits in with The Narrative. After all, Trump is an abuser, a womanizer, all things bad. If a woman says he did something a long time ago, even without substantiation, it is published and to be believed.

Yet instead of a journalistic discussion over the ethics of this mentality, a furor is raised over something else: the location in which the newspaper chose to place the story within its pages.

The outcry begins with anti-Trump activists accusing the *Times* of "downplaying" the rape article by publishing it on a subpage rather than the front page (or home page) on its website. Out for blood, these activists want the anti-Trump article to be moved to top billing on the front page. They know that prominent placement of any story in the *Times* prompts other news organizations around the world to repeat and amplify the story. To these activists, the purpose of news today isn't to inform, it is to influence. To establish The Narrative. To demonstrate whose side you are on. To show where you come down on a particular issue or controversy. To tell people what they should think.

In this way, a subnarrative comes into play. Trump critics want to convince influential institutions, such as the *New York Times* and other media, that they should not normalize President Trump by covering him normally. When they veer off script, they can expect to get called out.

So after the Trump rape allegations are published, critics take to social media with their complaints. *Times* executive editor Dean

Baquet quickly buckles. He moves the rape story to the front page and apologizes for not having given it higher billing in the first place. *News placement by popularity contest.*

"The fact that a well-known person was making a very public allegation against a sitting president 'should've compelled us to play it bigger,'" Baquet explains, as part of a mea culpa published in his newspaper. The *Times* also says, "Mr. Baquet said he had concluded that [the story] should have been presented more prominently, with a headline on The *Times*' home page." Also, Baquet states that the paper had been "overly cautious" by not putting the story on the front page to begin with.

Journalistically, "overly cautious" was exactly the right approach, given the circumstances. Baquet's new suggestion that the *Times* should amend its normal standards if an accuser is "well known" and the accused is the president is puzzling, to say the least. The responsibility to verify claims and follow guidelines does not fluctuate depending upon the relative notoriety of those involved. There is no more important time for a news organization to maintain its standards than when it comes to damaging allegations against prominent people.

But matters are about to become even more confused. Even as Baquet says he's made a mistake by not giving the story more prominence, an official statement from the *Times* indicates that, under its own news policy, the article should not have been published at all.

According to the *Times*, it had previously developed informal guidelines dictating how to treat sexual assault allegations like the ones against President Trump. The guidelines said a story would be only published if three conditions were met:

- The *Times* must locate sources outside those mentioned by the accuser.
- The additional sources must be able to corroborate the allegations.
- They must be willing to be named on the record.

The *Times* acknowledged that the story about President Trump had failed on all three counts. The newspaper was unable to find independent sources or any other additional corroboration of the accuser's story, and two women the alleged victim said would corroborate her story wouldn't allow their names to be published.

So the newspaper is simultaneously justifying and apologizing for not giving the story more prominence to begin with—while revealing it should never have been printed in the first place. This is a quintessential example of Orwellian doublethink: "To know and not to know, to be conscious of complete truthfulness while telling carefully constructed lies, to hold simultaneously two opinions which canceled out, knowing them to be contradictory and believing in both of them."

The whole snafu exposes the way The Narrative has managed to transform the way we in the news think about our own roles. Too many of us have fallen into the trap of believing it is our job to further a story line, rather than simply report facts and information. That is the reason observers occasionally ask me why I, too, have not joined the rest of the press to report a story or set of allegations. In their view, I should report the same "news" everyone else is reporting in order to show that I agree with or support it. They seem to find my response unexpected. I ask, *"Why are you so anxious for me to report something you already know and that has already been widely reported?"*

My goal is to bring information to light that is underreported or not well known. If someone wants me simply to report what others are reporting, they're not looking for facts and information. They want me to advance a narrative they support.

Columbia Journalism Review (*CJR*) fell into this trap with the Trump rape story. It published a blatantly slanted article arguing that in the *New York Times* and beyond, the story hadn't gotten the play it deserved. Incredibly, *CJR* turned to none other than the left-wing smear group Media Matters as a source of its evidence—while omitting the fact that Media Matters is a partisan outfit created by David Brock, an anti-Trump loyalist of Hillary Clinton. The organization has been supported by the liberal activist George Soros and other donors whose names are kept secret.

"As Media Matters for America's Katie Sullivan pointed out," *CJR* chided, the alleged victim's claim against Trump "did not make the front page of Saturday's *New York Times*, *Wall Street Journal*, *LA Times*, or *Chicago Tribune*; *The Washington Post* did put it on A1, but did not lead with it."

In taking this position, *CJR* put itself into lockstep with Media Matters and other propagandists who view the news as a tool to set agendas rather than disseminate facts. After all, nearly everyone reported the rape allegations. It's just that *CJR* and anti-Trump activists wanted the story to be even bigger and more ubiquitous. They aren't seeking information: they want a narrative to shape public opinion.

As a footnote, Carroll, President Trump's accuser in the story, gave an odd interview to CNN host Anderson Cooper. In it, she said that Trump's supposed assault was "not sexual" and added that rape is typically considered "sexy." CNN quickly cut to a commercial, but not before we heard the accuser tell Cooper in a low, sultry voice—leaning in and staring intently into his eyes—"You're *faaaaaascinating* to talk to."

The press provided no meaningful follow-up to the bizarre behavior by the accuser. In fact, shortly after Carroll appeared to be something less than credible, when her role in The Narrative was over, the national media instantly became uninterested in the formerly hot topic. It went down the memory hole as if it never happened.

And the *Times* is about to learn that making important publishing decisions based on social media pressure from activists simply ensures more pressure will come.

Headline Hullabaloo

The second big event at the *New York Times* in the summer of 2019 is a kerfuffle over what started out as a perfectly objective, accurate front-page headline on August 6 after several mass shootings: "Trump Urges Unity vs. Racism." Indeed, Trump had done just that: he'd urged unity over racism.

"In one voice, our nation must condemn racism, bigotry and white supremacy," Trump said. "These sinister ideologies must be defeated. Hate has no place in America."

But shortly after publication, several Democrats go public and argue that the *Times* should not have printed what Trump said. They claim the newspaper should have concluded that what Trump *said* was not what he actually *meant*. Or, they say, the newspaper should have chosen to highlight a gun control narrative instead.

In other words, though Trump's critics frequently complain that he does not do enough to promote unity and condemn racism, when he explicitly does so, they claim he doesn't mean it.

Presidential candidate Senator Cory Booker of New Jersey is one Democrat who goes public on Twitter to criticize the *Times*' entirely factual headline. "Lives literally depend on you doing better, NYT. Please do," he tweets.

Congresswoman Alexandria Ocasio-Cortez, a Democrat from New York, inserts race into the equation. She says the *Times* headline shows "how white supremacy is aided by—and often relies upon—the cowardice of mainstream institutions."

Again, executive editor Baquet surrenders to the critiques. He changes the headline to one that attacks Trump and advances a gun control narrative. "Assailing Hate but Not Guns" reads the replacement headline.

A top liberal-leaning executive at a national news organization commented on this turn at the *New York Times* when I asked him about the state of the media today. "The analytics they do in this era to find out who's clicking headlines and why has become so infected in their editorial decisions at the *New York Times* and other places. I look at the headlines, and I say, 'Holy shit! I know where they stand!' It's amazing how far they've gone down that path and that there aren't more objections to it internally."

It is worth inserting here that even as the *Times* is suffering internal trauma, the *Washington Post* is fighting its own demons. One prominent example is the *Post*'s headline fiasco in October 2019.

The *Post* headline tops a story about US Special Forces tracking down the man known as "al-Baghdadi," leader of the Islamic extremist terrorist group ISIS. US Special Forces cornered al-Baghdadi in Syria. He detonated a suicide vest, killing himself and three of his children.

The original *Post* Sunday headline reads, "Abu Bakr al-Baghdadi, Islamic State's 'Terrorist in Chief,' dies at 48."

Nothing much to argue about there. Or so one would think.

Not long after that headline is published, somebody at the newspaper mysteriously changes it. In the new headline, the *Post* miraculously—and shockingly—transforms al-Baghdadi from a "terrorist" into a "scholar." The new headline reads, "Abu Bakr al-Baghdadi, austere religious scholar at helm of Islamic State, dies at 48."

Besides the dubious nature of the headline, the article contains what some see as questionable references to the dastardly subject of the article. It refers to al-Baghdadi's supposedly soft-spoken manner, his scholarly wire-rimmed glasses, and his reportedly peaceful ways (before becoming the world's most wanted terrorist, that is).

The treatment of al-Baghdadi's obituary is so outrageous that it prompts a flurry of sarcastic responses. Twitter users make up ridiculously whitewashed obituaries of notorious figures under the hashtag #WaPoDeathNotices:

@robbysoave: Voldemort, austere political reformer and aspiring school-teacher, killed by teen terrorist. #WaPoDeathNotices

@thor_benson: Genghis Khan, noted traveler, dies at 64. #WaPoDeath-Notices

@KassyDillon: Hannibal Lecter, well-known forensic psychiatrist and food connoisseur dead at 81. #WaPoDeathNotices

@jason_howerton: Adolf Hitler, passionate community planner and dynamic public speaker, dies at 56. #WaPoDeathNotices

I tweet out my own contribution about the recently departed sex offender Jeffrey Epstein, who had reportedly strangled himself in prison:

Jeffrey Epstein, admirer and caretaker of young girls, dead at age 66, after coming down with a sore throat.

Amid the furor, the *Washington Post* decides to alter its headline a second time. This time, it reads, "Abu Bakr al-Baghdadi, extremist leader of Islamic State, dies at 48." A *Post* executive named Kristine Coratti Kelly tweets an apology for the earlier complimentary words about the ISIS leader but no explanation:

"Regarding our al-Baghdadi obituary, the headline should never have read that way and we changed it quickly."

Back to the *New York Times* saga. A third incident at the Gray Lady during the summer of 2019 comes on August 8, two days after the paper's Trump headline controversy. This one involves a nasty Twitter fight between *Times* deputy Washington editor Jonathan Weisman and a *Times* contributor named Roxane Gay. I read numerous accounts about the Twitter war and reviewed the operative tweets. They amount to inside baseball, and most Americans were unaware the squabble even took place. Yet it made news headlines in that inside-baseball kind of way we've grown accustomed to, where the national media in New York and DC talk to one another rather than inform those pesky news consumers in the rest of America.

Anyhow, the whole thing started when Weisman fired off a controversial tweet responding to the characterization of two members of Congress as midwesterners:

Saying @RashidaTlaib (D-Detroit) and @IlhanMN (D-Minneapolis) are from the Midwest is like saying @RepLloydDoggett (D-Austin) is from Texas or @repjohnlewis (D-Atlanta) is from the Deep South. C'mon.

The underlying idea is not that controversial: many red or purple states have blue cities that are nothing like the rest of the state politically. The problem came because some of the *Times'* most progressive readers perceived Weisman's comments to be racist and an insult to either the congresswomen or midwestern cities—or both. They thought Weisman was implying that some people are "real Americans" and some are not—a commonly debated narrative that Weisman wasn't specifically addressing. Suffice it to say that Weisman then became a target. Gay mocked him on Twitter. Weisman, in turn, apparently emailed her and demanded an "enormous apology." She then unleashed a profane, racially tinged response: "The audacity and entitlement of white men is fucking incredible." In the end, Weisman apologized for embarrassing the *Times*, and Baquet demoted him.

Later, a former *New York Times* staffer tells me that the event reflects serious hypocrisy. Newspaper management at the *Times* lets some employees run amok in the media but throws the book at others. "When it comes to policing social media postings and cable TV appearances, the *New York Times* has a double standard," the journalist says. "For the last four years *NYT*ers have been overtly anti-Trump in these forums. With a few exceptions, they have not been checked or sanctioned, though their behavior openly violates the paper's posted standards. Then when an editor posts a silly, stupid tweet that offended the leftist mob, the powers that be came crashing down on him."

The Meeting

As a painful August drags on, the next major event in the sad devolution of the *New York Times* happens when an insider leaks an audio recording of an embarrassing staff meeting convened on the heels of these public fiascos. The August 12 gathering reveals some of the *Times'* editorial staff to be endlessly self-reflecting yet utterly lacking in self-awareness.

The seventy-five-minute-long session is apparently called for the purpose of quelling discontent among vocal staff members who are demanding that the paper be pulled ever further to the left. The content of the meeting isn't meant to be shared with a public audience. But somebody provides a recording of it to the liberal website Slate, which publishes a "lightly condensed and edited transcript" (without explaining why it needed light condensing or editing). Even in its edited and condensed version, the transcript provides perhaps the most remarkable window ever offered into the crafting of international political narratives by a major news organization.

At the *Times* meeting, there is no debate over whether President Trump is a racist and a liar; it is simply just a matter of how the *Times* will convey how big of a racist and how bad of a liar he is and how often it will say so. The problem, as vocal staffers see things, is that the *New York Times* isn't treating Trump harshly enough to fulfill readers' expectations. The proverbial inmates appear to be running the asylum.

During the staff discussion, it becomes clear that the *Times* has predetermined what it sees as a primary narrative for the next two years ahead of any news events actually occurring. In doing so, *Times* management and staff tacitly acknowledge that they are in the business of slanting rather than reporting. Under their plan, real news will not be reflected objectively; it will be shaped to fit the preconceived narrative. Executive editor Baquet calls the *Times'* coverage of the recently disproven Trump-Russia conspiracy "Chapter One" of the "story" the newspaper has written on the president. Chapter Two, he says, will be a narrative focused on race.

He begins by addressing the unexpected turn taken when Special Counsel Robert Mueller recently concluded that neither Trump nor anyone on his campaign had conspired with Russia. "Did Donald Trump have untoward relationships with the Russians, and was there obstruction of justice? . . . We set ourselves up to cover that story. . . . And I think we covered that story better than anybody else," Baquet tells his staff. But he goes on to say, "The day Bob Mueller walked off that witness stand, two things happened. Our readers who want Donald Trump to go away suddenly thought, 'Holy shit, Bob Mueller is not going to do it.' . . . We were a little tiny bit flat-footed. I mean, that's what happens when a story looks a certain way for two years. Right?"

There is no hint of introspection about how the *Times* could have gone so far down the wrong trail and misread the tea leaves on its biggest story. *Are their journalists so blinded by bias that they missed the facts?* Nobody even asks the question.

Baquet also addresses the controversy over the "Trump Urges Unity vs. Racism" headline. He calls it "a fucking mess," the result of a "system breakdown." On the one hand, Baquet tells his staff, they shouldn't be reactive to the whims of Twitter. "Being independent also means not editing the New York Times for Twitter, which can be unforgiving and toxic." On the other hand, he portrays his newsroom as doing just that: editing for Twitter, rattled by the pushback from political figures on social media. "We were all over it, and then in the middle of it, [deputy managing editor] Rebecca Blumenstein sent an email—but we were already messing with [the headline]—saying, 'You should know, there's a social media firestorm over the headline.' My reaction [inaudible] was not polite. My reaction was to essentially say, 'Fuck 'em, we're already working on it.'"

The idea that a "social media firestorm" would send a serious news organization into a tailspin is reason for all of us to be concerned. It is the direct result of the phenomenon I wrote about in *The Smear*, where corporate and political interests manage to influence the news using strong-arm propaganda tactics. They know they need only take up enough bandwidth on Twitter (often using AstroTurf methods, robotics, and even fake accounts) to be able to influence public policy and—as we now see—opinion at one of the most important news organizations in the world.

As the *Times* meeting continues, a staffer complains that his own personal viewpoint wasn't reflected in the original "Trump Urges Unity vs. Racism" headline. "The headline represented utter denial, unawareness of what we can all observe with our eyes and ears. . . . A headline like that simply amplifies without critique the desired narrative of the most powerful figure in the country," the employee grumbles, according to the transcript. "If the Times' mission is now to take at face value and simply repeat the claims of the powerful, that's news to me. I'm not sure the Times leadership appreciates the damage it does to our reputation and standing when we fail to call things like they are." In other words, the speaker seems to believe that narratives are to be avoided, except when they reflect his or the *Times'* desired narratives, such as "Trump is a racist." When Trump says something contrary to the racist narrative, he is to be contradicted and redefined. I'm not sure the complaining *Times* staffer appreciates the damage done to the newspaper's reputation by such journalistic failures.

To the uninitiated, the discussions exposed in the transcript may be shocking. But to insiders, it is business as usual. It doesn't seem to occur to the journalists who are part of the conversation that they are demonstrating a serious lack of objectivity, undercutting the idea that they can cover their most important story with the fair and fact-based neutrality it requires.

Many journalists today misunderstand the role of a reporter in fact-checking and holding the powerful accountable. If a company says it produces more microwaves than anyone else and there's evidence it doesn't, it's important to include that contradictory information. If a political leader claims that seventy senators voted for a bill but only sixty-eight did, the record should be corrected. Those are facts. But once you deviate into the territory of calling a speech "heartfelt" or "uninspiring" or claim to know whether voter fraud is or isn't likely to occur in the future, you're no longer writing news; you're giving your opinion or making personal predictions about the future. That sort of material belongs in opinion columns.

Back to the meeting at the *New York Times*. The transcript does reflect a bit of internal pushback to the notion that the *Times* and its headlines go too easy on Trump. Associate Managing Editor for Standards Philip Corbett takes a fairly brave stance (considering his audience): "I would dispute the idea that when we have made mistakes about headlines in the last months or couple of years that they have always been in the same direction . . . that the mistakes you're seeing are when we're going, shall we say, too easy on Donald Trump. There certainly have been headlines where I feel like that has been a failing. But I will say, honestly, there have been headlines that many of us have been concerned about or asked to have changed or have had discussion about where I felt the problem was the opposite, where we were showing what could be read as bias against Trump, and were perhaps going too far in the opposite direction."

But the far greater sentiment at the meeting is viciously anti-Trump. And it is against this backdrop that the *Times* codifies a commitment to forwarding the Trump-as-racist narrative over the next two years.

"Race in the next year—and I think this is, to be frank, what I would hope you come away from this discussion with—race in the next year is going to be a huge part of the American story," Baquet tells the staff. "And I mean, race in terms of not only African Americans and their relationship with Donald Trump, but Latinos and immigration." He goes on to ask, "How do we cover America, that's become so divided by Donald Trump? . . . You all are going to have to help us shape that vision. But I think that's what we're going to have to do for the rest of the next two years. . . . [T]his is a different story now. This is a story that's going to call on different muscles for us. The next few weeks, we're gonna have to figure out what those muscles are."

It seems to me there are countless fair and rational questions that honest *Times* staffers could have asked at this point. Americans may be divided over Donald Trump, but did Trump alone *really* cause the divisions? Or did the media coverage and his enemies stoke that dynamic? Are most Americans coexisting just fine when it comes to race and other media hot points, with the divided version existing in amplified renditions on social media, on the news, and in New York, DC, and LA? Somebody in the room might have offered, "A great number of Americans aren't racist and don't believe their president is. We should work hard to try to understand and represent their views in some of our stories—be balanced and open minded." A staffer could even have said, "Look, we may all hate President Trump—I know I do—but our personal feelings and political beliefs shouldn't impact our news coverage. That belongs on our editorial page but not in our news reports."

Instead, *Times* staffers ask management questions such as "Could you explain your decision not to more regularly use the word *racist* in reference to the president's actions?" And here's another staff comment with tremendous implications: "I'm wondering to what extent you think that the fact of racism and white supremacy being sort of the foundation of this country should play into our reporting. Just because it feels to me like it should be a starting point, you know? Like these conversations about what is racist, what isn't racist. I just feel like racism is in everything. It should be considered in our science reporting, in our culture reporting, in our national reporting. And so, to me, it's less about the individual instances of racism, and sort of how we're thinking about racism and white supremacy as the foundation of all of the systems in the country."

It's certainly reasonable for reporters to explore subjects or trends that haven't been well covered in the past because they may not have been as visible to whites. But if you start off on the front end planning to slant every story to fit The Narrative that "America is racist," you're not functioning as a news reporter; you're the one with the bias issue.

Of course, not everyone at the *New York Times* approaches news in this slanted way. The newspaper still publishes some great reporting by outstanding journalists. But who among them would feel comfortable speaking out with an alternate viewpoint in an environment like the one reflected by the meeting transcript?

After the transcript was made public, a former longtime *New York*

Times staffer who read it tells me, "The media are increasingly reliant on narrowly targeted audiences, both financially and psychologically. The *NYT* subscriber base is overwhelmingly anti-Trump. This impacts the reporting priorities, the choice of 'experts' quoted in stories, the universe of anonymous sources utilized, the tone and shaping of stories, their headlines, and finally, their placement. The viral nature of social media reinforces all these tendencies. Of course, there are exceptions, but this is the overall trend. The money is with the niche audience. That requires giving up your news values."

When Airplanes Attack

The next notable event in the public unwinding of the *New York Times* in the summer of 2019 comes on September 11. The *Times* publishes a controversial, tone-deaf tweet that seems to assign responsibility for the September 11, 2001, terrorist attacks to airplanes rather than foreign terrorists:

> 18 years have passed since airplanes took aim and brought down the World Trade Center.

By now, the *Times'* cycle of publication, public pushback, and correction have become almost routine. After some high-profile social media outrage, the *Times* deletes the original tweet with the following note:

> We've deleted an earlier tweet to this story and have edited for clarity. The story has also been updated.

The *Times'* new and improved replacement tweet removes "airplanes" as the attackers but still fails to mention the true perpetrators: Islamic extremist terrorists. The new tweet reads:

18 years after nearly 3,000 people were lost, families of those killed in
the terror attacks will gather at the 9/11 memorial. There will be a moment of silence at 8:46 a.m., then the names of the dead—one by one—will
be recited.

As the shadows grow longer, the leaves began to turn, and a crisp
note fills the air, there is still time for the *New York Times* to wrap itself
in one more scandal in the summer of 2019. On Saturday, September 14, the newspaper publishes a story that again puts its reporting
and ethics under the microscope.

This article accuses Supreme Court Justice Brett Kavanaugh of
sexual misconduct at a dorm party decades earlier. The report is
bylined by the *Times'* cultural reporter Robin Pogrebin and her colleague Kate Kelly to promote a new book they'd written, *The Education of Brett Kavanaugh: An Investigation*. In the book and the article, a
former Kavanaugh classmate named Max Stier claims he observed
Kavanaugh exposing himself and forcing a female classmate to touch
his penis. Democrats running for president, including Senators Kamala Harris and Elizabeth Warren, immediately cry for Kavanaugh's
impeachment.

But there are serious problems with the article. First and fore-
most, it failed to mention that, incredibly, nobody from the *Times*
ever spoke to the supposed victim. And it omitted another even more
crucial fact: the supposed victim does not remember the incident
happening at all. *In other words, there is no verifiable victim of the alleged
sexual assault.*

Obviously, under no journalistic standard should such a story
have been published in the first place.

All of this is uncovered only after the *Times* article is published,
when an outside reporter obtains a copy of the book being promoted.
In the book, the writers disclose that the "victim" has no recollection of Kavanaugh exposing himself or doing anything worse. The
reporting lapse turns out to be so egregious that the *Times* finds itself
in the unusual position of being taken to task by liberal and conser-

vative media alike. Journalist David French writes in *National Review* that it is unconscionable for the *Times* to have left out the part about the alleged victim having no memory of any assault. "All in all, the story was one of the worst examples I've ever seen of neglecting story for narrative," French says. "The true story casts strong doubt on the narrative that many *New York Times* readers and staffers firmly believe; so the *Times* fed its readers the narrative."

Even the left-leaning CNN notes, "The New York Times was reeling on Monday after its Opinion section fumbled a high-profile story about an allegation of sexual misconduct against Supreme Court Justice Brett Kavanaugh, drawing widespread criticism and condemnation of the newspaper. It was the latest in a series of high-profile blunders that has caused embarrassment to James Bennet since he was appointed in 2016 as the editor overseeing The Times' Opinion section."

Scott Shapiro, a liberal law professor at Yale, calls the *Times'* article "outrageous" and tweets, "Would love to see my fellow liberals who routinely threaten to unsubscribe to the NYT make the same threat now." Over at the liberal NPR, David Folkenflik makes the strange argument that journalists do not necessarily need an actual victim's memory or account at all to allege an attack. He tweets, "One can argue that the [alleged victim's] failure to remember [any sexual incident], given her intoxication, is not dispositive." But even he parts from the *Times*, adding "One can't argue, however, that that fact didn't need to be in the Kavanaugh story from the outset."

Facing a backlash worse than all the other recent ones, the *Times* publishes a correction in the form of an editor's note that reads, "An earlier version of this article, which was adapted from a forthcoming book, did not include one element of the book's account regarding an assertion by a Yale classmate. . . . The book reports that the female student declined to be interviewed, and friends say that she does not recall the incident. That information has been added to the article."

In an opinion piece published by *USA Today*, Paul Janensch, a former newspaper editor who taught journalism at Quinnipiac University, says he is unsatisfied with the *Times'* correction. "The 'earlier version of this article' never should have appeared in the *New York Times*," he asserts.

In subsequent interviews, the *Times* reporters explained their omission by stating "During the editing process there was an oversight and this key detail, about the fact that the woman herself has told friends she doesn't remember it and has not wanted to talk about it, got cut and it was an oversight and the *Times* adjusted it and we're very sorry that it happened."

But still more issues arise with the article.

It turns out the *Times'* supposed corroboration of the sexual assault claim was no corroboration at all, in any journalistic sense. The reporters did not talk with anyone who heard the victim speak of the assault or who witnessed it themselves. Their verification consisted of talking to two other people who'd heard the "eyewitness," Stier, tell the same story. And the *Times* forgot to disclose Stier's political ties. *Times* reporters Pogrebin and Kelly described him as someone who "runs a nonprofit organization in Washington." In their book, they referred to him as "a respected thought leader on federal government management issues in Washington, as the founding president and chief executive of the Partnership for Public Service." But they left out an important part: the part about Stier having worked on Bill Clinton's impeachment defense team.

In a later interview on The Hill webcast *Rising*, host Saagar Enjeti presses the *Times* reporters about their selective descriptions of Stier: "As I understand it, he was on the opposing team from Kavanaugh during the Clinton impeachment and his wife was denied a federal judgeship by the GOP. Did you include that information in your book? I mean, that seems like pretty clear evidence of a vendetta against Brett Kavanaugh."

"You know," replies Pogrebin, "we didn't include it in the book—we do talk about what he's been doing for most of his career, which is nonpartisan." She seems to be admitting they made a conscious decision to cast Stier as nonpartisan.

"You don't think that's germane?" interrupts Enjeti. "I mean, for somebody to accuse somebody like this . . ."

"Well, I mean . . . Is it germane?" Pogrebin replies, repeating Enjeti's question.

". . . and the victim does not even remember this incident, you don't think that is germane detail?" Enjeti persists.

Instead of answering, Pogrebin fires back. "Do you think it's germane that Brett Kavanaugh wrote the Starr Report [in the impeachment of President Clinton]?"

"Yes, absolutely!" replies Enjeti, going on to point out that the fact was widely reported by the *Times* and other media and that Kavanaugh was questioned extensively about it in his confirmation hearing—unlike the case with Stier, whose political ties were not addressed or even mentioned by the *Times* reporters.

A third issue with the *Times* story arises with an odd social media post written by Pogrebin and tweeted from the official *New York Times* opinion page Twitter account. It is about another Kavanaugh accuser, Deborah Ramirez:

Having a penis thrust in your face at a drunken dorm party may seem like harmless fun. But when Brett Kavanaugh did it to her, Deborah Ramirez says, it confirmed that she didn't belong at Yale in the first place.

That prompts expressions of outrage over the cavalier use of the description "harmless fun." The *Times* deletes the tweet.

A fourth question comes up several days later. In interviews promoting their book and defending their reporting, the *Times* reporters acknowledge that Kavanaugh had agreed to talk to them, apparently on background or off the record, but that they refused those terms. The problem with that is they agreed to let Kavanaugh's accusers and opponents speak anonymously or without being quoted. In terms of ethical and fair treatment, it's obviously a problem if terms or courtesies are offered to one side but denied to the main target of an accusation.

No matter. As the primary allegations in the reporting fall apart, and even with all the pushback, the narrative train speeds down the track, impossible to stop. The story's many problems are deposited neatly down the memory hole as if they never happened, as far as some Democrats are concerned. The billionaire environmental activist Tom Steyer publicly calls for Kavanaugh's impeachment, tweeting:

> The @GOP is so hell bent on guaranteeing a conservative court, they are willing to overlook serious allegations of sexual misconduct and perjury. The system is broken.

Senator Kamala Harris keeps her original tweet pinned to her home page for days after the *Times* corrected its article and admitted there was no verified victim. The tweet reads:

> Brett Kavanaugh lied to the U.S. Senate and most importantly to the American people. He was put on the Court through a sham process and his place on the Court is an insult to the pursuit of truth and justice. He must be impeached.

Other slanted writers and reporters twist themselves into pretzels looking for ways to use the *Times'* story to advance their own anti-Kavanaugh narratives. A writer named Alicia Cohn mentions the *Times'* correction only as an excuse to regurgitate earlier unproven and discredited smears against Kavanaugh. "[Kavanaugh] denied a previous claim of sexual assault last year during his confirmation hearings for Supreme Court justice," she writes in The Hill. "In that incident, the woman involved, Christine Blasey Ford, testified before Congress about the allegation." Cohn omits—much as the *Times* omitted—the fact that Ford's account was unsupported. Other alleged witnesses denied remembering any such event. That includes Ford's own friend Leland Keyser, who said, "I don't have any confidence in [Ford's] story" and said she doubts the incident ever happened.

Not only does Cohn leave all of that out of her article, she also

continues the regurgitation of unsubstantiated allegations, reporting that "Another woman, Deborah Ramirez, last year accused Kavanaugh of exposing himself and forcing her to touch him during their
freshman year at Yale during the 1983–84 school year." Cohn fails
to mention that none of the alleged witnesses in that case corroborated Ramirez's claim, either. Cohn continues, "A third woman,
Julie Swetnick, accused Kavanaugh of being part of a group that
planned gang rapes of young women at house parties while in high
school." Cohn leaves out that Swetnick later backtracked and contradicted her sworn statements during an infamous interview with
NBC News. Cohn also chooses not to disclose that Swetnick and her
attorney, Trump foe Michael Avenatti, were referred to the FBI for
investigation over their "subsequent contradictions . . . the lack of
substantiating or corroborating evidence, and the overarching and
serious credibility problems pervading the presentation of these allegations."

Avenatti was later arrested for alleged financial crimes, extortion,
and cheating his clients.

Firing the Public Editor

I can't help but think that the angst-filled newsroom at the *New York
Times* might not have to expend so much effort dodging flak if management had allowed the paper's public editor to do her job. The public editor was the internal ombudsman assigned "to help keep the
Times and its coverage honest in an increasingly commercialized
and politicized news environment." This was the person assigned
to address major public criticism and, to some degree, inoculate the
newsroom from having to get mired so deeply in controversies over
its coverage.

The position of public editor at the *Times* was first created after
the Jayson Blair scandal. Blair was the *Times* reporter who resigned

in disgrace in 2003 after it was discovered that his stories—some of them published on the front page—were fabricated and plagiarized. The controversy led to the resignation of *Times* executive editor Howell Raines and managing editor Gerald Boyd. The new public editor would serve as a check and balance to help uncover and remedy journalistic misdeeds sooner.

In May 2016, Elizabeth Spayd became the *Times*' last public editor. During her relatively short tenure, she fielded criticism about controversies such as the *Times*' increase in "native advertising," meaning ads seamlessly worked into the fabric of the publication as if they were a news story. Spayd called the uncomfortable mix of commercials and journalism a proven winner in terms of revenue. She noted that "The vast majority of readers apparently find it unobjectionable." She drew that conclusion in part, she said, because she had received few complaints about it. Actually, the lack of complaints might have been because most readers don't recognize native advertising when they are reading it. That's the whole point: it is advertising disguised as news.

In any event, during the course of her work, Spayd sometimes criticized her own publication. In turn, she was sometimes criticized by *Times* staffers and outside journalists. That comes with the territory. Some of Spayd's critics took their objections to left-leaning outlets such as *The Atlantic*, which worked to controversialize and undermine her. *The Atlantic* printed accusations of her being "inclined to write what she doesn't know," said her work had become "iconic in its uselessness and self-parody," and accused her of "squandering the most important watchdog job in journalism."

In May 2017, *Times* publisher Arthur Ochs Sulzberger, Jr., suddenly eliminated the job of public editor. In a memo explaining Spayd's termination, he argued the position was now superfluous because the Internet had become the media's watchdog. "Our followers on social media and our readers across the Internet have come together to collectively serve as a modern watchdog, more vigilant and forceful than one person could ever be," he stated. "Our responsibility is to empower all of those watchdogs, and to listen to them, rather than to channel their voice through a single office."

So as the 2020 presidential campaign heated up, the *Times'* disastrous summer of 2019 drew to a merciful close. If, as you read these words, you remember a dominant narrative on the news and the Internet being President Trump as a racist and divisive leader in a divided America, you know it is at least partly the result of a plan executed by the *New York Times*. Never did Trump's slur, "the failing *New York Times*," seem to hit closer to home.

The New "Woke" Times

If there were any hope that the cacophony of high-profile embarrassments at the *Times* would prompt a rational reexamination of the newspaper's slant, it was not to be. The exclamation point was added in June 2020 in what became perhaps the most bizarre of the incidents. Once again, the media finds itself giving time and space to news about the news rather than, well, to news itself.

The trigger of this disaster is an op-ed written for the *Times* by Senator Tom Cotton, an Arkansas Republican, titled "Send in the Troops." It advocated for dispatching military assistance to US cities wracked by violence amid protests against police brutality. Sending in the troops is a position that has plenty of both supporters and detractors. This is typical of topics tackled in op-ed pieces. Most people would argue that's the point: to present diverse views.

But that's not how a lot of *Times* staffers apparently see things. After the piece is published, they launch an internal revolt and take to social media to denounce their own newspaper for having dared to publish Cotton's words. "Running this puts black @NYTimes staff in danger," claim staffers and their supporters in unified tweets.

Then *New York Times* employees pen a letter to *Times* management demanding the newspaper publish a refutation of Cotton's position.

The letter is addressed to editorial page editor James Bennet and his two deputies, chief executive Mark Thompson, chief operating officer Meredith Kopit Levien, executive editor Baquet, and publisher A. G. Sulzberger, who by then had taken over the newspaper's helm from his father. In the fallout, Bennet admits that he didn't read the Cotton op-ed before it was published and left that job up to his number two.

As demanded by the mob of staff, the *Times* adds an editor's note to Cotton's column. It says the op-ed had been approved in a "rushed" editorial process that did not meet its standards. Sulzberger emails employees, "Last week we saw a significant breakdown in our editing processes, not the first we've experienced in recent years."

Days later, editorial chief Bennet is finished. The *Times* announces he has resigned. The same newspaper that defended controversial op-eds such as the one signed by Russian president Vladimir Putin in 2013, and even an anti-Trump opinion piece that hadn't been signed at all and was published anonymously, was now adjudicating the words of a prominent US senator to be just too incendiary.

It is a strange place, indeed, where news reporters can editorialize but op-ed editorials cannot. On June 1, 2020, after President Trump walked through a public Washington, DC, park amid national protests and riots—to a church that had been burned a block away—*Times* chief White House correspondent Peter Baker, tweets:

> Trump just stands in front of the church and holds up a bible while posing for photos. He does not even go inside for a faux tour of the damage or make a pretense of having any purpose in going there other than to pose for photos.

The latest developments remove any lingering doubt as to how the *Times* sees its modern mission: serving and pleasing the left-wing activists on its staff and the liberal activists who dominate on the news and social media. Arthur Ochs Sulzberger's dictum when he fired the newspaper's public editor in May 2017 had come to pass in a terrible way. Recall that, at the time, he declared that the *Times*' followers on social media would "collectively serve as a modern watchdog, more vigilant and forceful than one person could ever be."

The *Times* let itself become hopelessly slanted. Captive to organized feedback on social media. Beholden to irredeemably conflicted staff members. Consumed by internal demons.

Make no mistake: other media outlets are taking note. In this way, they are motivated to self-censor news and information, lest they draw the wrath of the mobs. One editorial figure at a major international publication who did not want to be identified recounted numerous pieces he has recently killed for fear of the organized backlash.

"They can bankrupt me," he tells me. "Facebook, Twitter, Google—they can ruin you in a matter of hours. For somebody like us, they can destroy you. So what do we do? We pull our punches. To raise certain issues is to cut your own throat." He continues, "The newsman in me says, 'Tell the truth,' and that sounds great. But if I do that and destroy [my publication] in the process, what kind of pyrrhic victory is that?"

The information landscape becomes ever narrower, squashing diversity of thought and facts. Pretty soon, we won't know what we don't know. And that will be that.

One former insider wrote to me after Bennet's resignation and referred to his alma mater as "The New Woke Times." "The *NYT* still plays an outsized role in shaping the news agenda in America," he tells me. "The news business is highly competitive. If the leader has abandoned its own standards and history of fair and rigorous reporting, why should the others be any different?"

CHAPTER 6

The Verbiage of The Narrative

LIES, EVIDENCE, AND BOMBSHELLS

I'm on assignment in Raleigh, North Carolina, swapping stories in a local pub with colleagues who also work in national news. We see a news program on a TV monitor over the bar. A banner at the bottom of the TV screen claims that whatever the hosts are talking about is "a bombshell!" We comment on how it seems as if every few days, news reporters are characterizing some piece of information as a "bombshell." Life these days is full of bombshells, most of them, ironically, unremarkable. It is a term I rarely used on the news and seldom heard other reporters use until the last couple of years. Now most anything and everything amounts to a bombshell when it is something that the media want you to believe is true or important—even if it's not.

"The thing is," I remark to my media friends, "if something really is a 'bombshell,' there's usually no need to call it that because viewers would know all by themselves. You wouldn't have to tell them." My colleagues concur. One of them, who has worked at two national news networks, offers a theory: "I think 'bombshell' is a natural outgrowth of the compulsion of news organizations to label everything as 'breaking news.' If everything is 'urgent, breaking news,' including ordinary events, how do we distinguish that from the stuff that really *is* sort of breaking news? 'Bombshell!'"

The rest of us nod in agreement. He continues, "So 'breaking news' is now the starting point for regular, ordinary news. Then, if there's something more than ordinary, it has to be bigger than 'breaking news,' so it's a 'bombshell.' I'm quite sure some other term will emerge when we realize 'bombshell' has been overused so much it's lost its meaning," he says, becoming more animated. "We'll call it 'supercalifragilistic news'!"

There's no doubt there are terminology trends in the news that can be directly traced to narratives. This chapter will explore several as they relate to narratives about Donald Trump.

Why does this chapter focus on The Donald? Because he is the biggest reason we have arrived where we are in journalism today. For several years, he has been, by far, the single most frequently covered topic on national news. The press has used Trump's perceived flaws and vulnerabilities as an excuse to justify very unjournalistic language and behavior. *"After all,"* they reason, *"he's a liar and deserves poor treatment."* Thus there would be a gaping hole in this book if we failed to examine his critical place in the dynamic.

I find the news media's transformation of their traditional role when covering Trump to be particularly dangerous to the larger pursuit of facts and to media credibility. Certainly, members of the public are free to judge Trump or anyone else however they wish. But we in the press have a different role and responsibility. No matter how we feel about Trump or any other subject of our reporting, we are not entitled to exaggerate about them, publish poorly sourced reporting, or treat them unfairly under the rationale that they somehow deserve it.

Trump is the vehicle that the media at large has used to unleash its furor and redefine journalism in a way it was never defined before. This is why the dominant news coverage claimed Trump was colluding with Russia, but that proved to be false. Few reporters covered this crucial story fully and fairly. After all, they reasoned, Trump is a liar, and the allegations against him are too good to pass up or fully vet. If the media wish to label him a Russian spy with no evidence, so be it. Why should he get a fair shake?

Once reporters got away with publishing sloppy, slanted, and opinionated journalism against Trump and Pulitzers were awarded, they started taking further journalistic liberties. Pretty soon, the Trump treatment began to bleed over into other aspects of news coverage. It has chipped away at the reputation of the media at large.

Alberto Martínez, the liberal University of Texas professor who documents narrative trends, opposed Trump in 2016. But he is objective enough to find fault with what he sees as the media's rampant narratives about Trump, which have, in turn, eroded public trust in the press. I asked him point-blank whether he thinks the media's declining credibility is Trump's fault—or their own. He answered by recounting how radically the media's description of Trump changed when he entered the race for president.

Martínez tells me, "In 2004, a CNN documentary described Trump as 'beloved,' 'the world's most popular businessman,' 'literally the gold standard,' 'a national phenomenon,' 'Trump has always worked hard and lived clean,' and 'He really is very smart, very sassy, very tough, but a warm and caring guy.'" He goes on to note that by 2016, however, news pundits portrayed candidate Trump as "the epitome of reckless ineptitude, fraud, failure, bankruptcy, bullying (who even mocked the disabled), vulgarity, adultery, homophobia, sexism, blatant racism, xenophobia, Islamophobia, authoritarianism, fascism, criminal sexual aggression, and dangerous warmongering. This extraordinary transformation happened because pundits constantly chose to interpret Trump's casual, careless words in the most horrifying ways imaginable. Even if two, or five, or ten of these stories about Trump were true, how could they all be true? I think the media lost much credibility because of its very partisan and relentless penchant to exaggerate and demonize."

As the conservative radio host Chris Plante wryly puts it, Trump's enemies take him literally while claiming he can't be taken seriously.

Martínez goes on to take apart what he calls a major media narrative about candidate Trump: "that he would be dangerous for minorities, especially Hispanics, black people, and Muslims. Countless news stories clanged alarms about this on CNN, MSNBC, the *New York Times*, the *Washington Post*, et cetera. If such news were accurate predictors of voters' concerns, it would mean that Trump pandered to white voters (especially those concerned about illegal immigration) and especially white supremacists, at the expense of offending minorities and losing their votes. But surprisingly, Trump's actual effect on voters was the very opposite. By comparing him with the previous Republican presidential candidate, Mitt Romney, I found that Trump won more votes from Hispanics, more votes from African Americans, and more votes from Muslims. Moreover, Trump won fewer votes from white people! Yet nearly all the media obscured this, because it didn't fit the narrative. On CNN, for example, Van Jones exclaimed that Trump's election was 'a white-lash against a black president!'"

In the end, this kind of reporting damages the media's reputation, and the public gets the short end of the stick.

There is a reason journalism standards exist in the first place—not just to afford fair treatment to people we like. They are also supposed to ensure fairness and accuracy when we cover those whom we don't like, don't agree with, or even believe are liars. In fact, that is when our standards matter most. It is a little bit like free speech: non-objectionable speech seldom needs defending. It is the difficult and controversial speech that demands free-speech protections. Trump tested our ability to prove how committed we are to staying true to our mission of journalists. And we failed.

The special verbiage the media deploys against Trump like nobody before him includes two key phrases: "lies" and "without evidence."

Lies

Before the era of Trump, we news reporters might have pointed out "discrepancies" or "contradictions" between two claims made by a newsmaker. We might have noted that a statement or claim was "disputed" or had proven to be "incorrect" or even "false." But we did not declare newsmakers' statements to be "lies."

I have countless professional examples of my own. As a reporter for about forty years, I don't recall ever calling anyone a "liar" in my news stories, though I suspect I've run into a fair number of lies.

In 1999, the FBI claimed that an accused Chinese spy, a Taiwanese scientist at the Los Alamos National Laboratory named Wen Ho Lee, failed a polygraph test. As a CBS News reporter, I obtained the lie detector test results and learned that the FBI had framed him. But I didn't accuse FBI officials of "lying." I reported that Lee passed the polygraph with flying colors, contrary to the FBI's claims.

In 2000, when I began breaking news about deadly rollovers of Ford Explorers equipped with Firestone tires, it was international news. The companies repeatedly insisted that no data or tests had predicted the disastrous combination of the faulty Firestone tires on a Ford vehicle that rolled over too easily during a tire blowout. Their claims were provably false. I obtained unequivocal documentation, testimony, and test results showing concerns had been repeatedly flagged years before. However, I never reported that Ford or Firestone executives had "lied"; I reported that their current statements were contradicted by their own documents.

When I covered Red Cross fraud after the September 11, 2001, Islamic extremist terrorist attacks, top officials at the charity insisted there had been no internal wrongdoing. I obtained internal audits that flagged widespread theft and misconduct involving Red Cross donations. But I didn't say in my news reports that Red Cross officials were "lying"; I simply pointed out the contradictions.

When I interviewed Republican congressman Stephen Buyer in 2009 about his questionable "charity," which had actually been funded by pharmaceutical and tobacco interests around the time he introduced legislation or gave speeches that benefited them, I pointed out that the record directly contradicted his claims. But I did not call him a "liar."

And in 2013, after President Obama told Americans they would be able to keep their own health care plans and doctors under the Affordable Care Act, or "Obamacare," I obtained internal documents showing his administration had forecast in advance that would not be true for millions of people. The journalism nonprofit Poynter Institute's fact-check feature, Politifact, called Obama's claim the 2013 "Lie of the Year." Still, I didn't call Obama a "liar." I simply pointed out that his own analysts predicted the opposite of what he'd claimed.

Why avoid the use of the "L" word in news reporting?

A lie is a very specific thing and, short of a confession, requires a reporter to claim to know the mind of the person who is supposedly lying. In fact, when someone gives seemingly contradictory information or makes a false statement, it could be for other reasons. The Ford and Firestone executives who made incorrect statements in 2000 about the safety of tires could argue they never saw the documentation from years past. It is possible their staff gave them poor briefings or withheld the information. Or they might have seen the documents but misinterpreted them or forgotten what they said. As unlikely as those excuses may seem, it is not the place of a news journalist to claim to know what is in a person's mind. There are very few instances I can think of where it is appropriate for a reporter to claim that a newsmaker "lied."

For me, this idea was put to the most challenging test with Hillary Clinton's false claim in 2008 that she had dodged sniper fire in Bosnia on a trip as first lady in 1996. I happened to know it was not true because I was with her on that trip as a reporter for CBS News. We most definitely had not been shot at. I even had the video from 1996 to prove it. (Clinton eventually retracted her statement and apologized.) But in my reporting on the controversy, I didn't call Clinton's statements "lies," nor did I refer to her as a "liar." There were other possibilities, however implausible they may seem. Technically, she could really have believed in her own mind that we *had* been shot at.

Perhaps she told herself a story so many times over the years that she had come to believe it was true. Maybe she had lost touch with reality. Whatever the case, as much as it may have seemed that she was lying, it was not my place to state on the news that I knew what was inside her head.

There is another important reason to be circumspect, as a news journalist, with accusing people of lies. To the news consumer, such language tends to sound as if the reporter is being biased or pejorative. It begins to feel very personal. It removes the sense of neutrality we try to maintain in order to be seen as fair reporters of fact. It is best to stick to the facts and let members of the public form their own conclusions.

The practice I have described, the reluctance to pretend to see inside the mind of subjects we report on, used to be considered the norm in journalism. But all of that has gone out the window. Now reporters frequently take sides, boldly declaring one newsmaker to be telling the truth and another (usually Trump) to be "lying"—even when the truth is unproven or impossible to know; when the differences are matters of interpretation or clashing opinions; or when the "lies" are actually exaggerations or misstatements. The *New York Times* has led the way in declaring itself arbiters of Trump's "lies." As we have seen, other media quickly followed suit, and biased observers heralded it all as brave and groundbreaking!

Just eighteen months into the Trump presidency, I conducted a Google search using the term "Obama lies." It produced 56 million results. I then searched using the term "Trump lies." That returned 630 million results. In other words, after only a year and a half in office, Trump got eleven times more search returns about his "lies" than Obama had after eight years in office. A closer analysis shows that many of Trump's Internet hits are links to news stories and blogs that list his "lies." In contrast, with Obama, many of his hits are articles or blogs defending him against accusations of lies—or with Obama officials accusing Trump of lying.

Does Trump "Lie" More Often?

It is a fair question to ask whether "Trump is the biggest liar ever" is a false narrative or rooted in truth. I am not sure there is an objective way to know. One reason is that the measure of Trump's—or anyone's—lies can often be a subjective exercise, as I have already discussed. It requires knowing, not just assuming or deducing, a person's knowledge and intent. Another reason it is so difficult to divine the truth is that some in the media set out to tarnish and destroy Trump in a way they have never done with any other political figure.

This is best exemplified by an interview I conducted with a former editor of the left-leaning website Politico, Susan Glasser. In 2017, I was speaking with her for a story about "fake news." Although I had not asked about Trump in particular, Glasser frequently inserted criticisms of Trump into her answers. In one instance, she told me that Politico had assigned a team of reporters to "fact-check every word out of Donald Trump's mouth" for an entire week of the 2016 campaign. She said her team had discovered that "Donald Trump uttered a lie or an exaggeration or a falsehood once every five minutes."

For comparison, I asked Glasser the obvious question: *What was the result of Politico's fact-check of Trump's opponent, Hillary Clinton? What was her lie-per-minute rate?* To my surprise, Glasser replied that Politico didn't have the resources to fact-check Hillary, too.

I didn't challenge Glasser on that inequity because that wasn't the topic of the interview I was conducting. She was graciously giving me her thoughts for a story about fake-news trends. But the surprise I felt at her response must have shown on my face. I can't think of any reputable journalism outfit, prior to the era of Trump, that would have considered fact-checking one and only one presidential candidate to be a legitimate news endeavor. But as far as I can tell, nobody else blinked an eye at this practice.

I have mentioned that it was on September 17, 2016, prior to the election, that the *New York Times* first branded Trump a "liar" on its front page. That particular article exemplifies the problems created by journalists claiming to know what is in a man's mind. The story criticized Trump for having questioned whether President Obama was born in the United States or in Kenya, where Obama's father was born. Obama supporters could certainly argue that Trump was wrong to question Obama's birthplace. But it didn't necessarily amount to "lying," particularly if Trump believed it to be a legitimate question. Obama had declined to produce his birth certificate during several years of speculation and ultimately, in 2011, released a certified copy of a birth certificate indicating he had been born in Hawaii. Most people accepted that as authentic proof of Obama's birthplace. However, since it was not an original birth certificate, and since it was released by the White House rather than the Hawaii department of vital records, and since there were no public eyewitnesses to Obama's birth in Hawaii stepping forward, it is not our place as journalists to unequivocally state that Trump was "lying" when he pursued the question. That can be an unpopular and difficult fact for some to accept because their personal feelings and biases overwhelm their analytical self.

New York Times executive editor Dean Baquet was among those who stepped away from factual analysis. He said it would "almost be illiterate [*sic*] to have not called [Trump's challenge of Obama's birthplace] a lie."

Loving the "Lies" Label

Instead of providing critical pushback to this questionable journalistic shift, in which reporters call opinions of their enemies or opinions they disagree with "lies," *Columbia Journalism Review (CJR)* jumped on the bandwagon, patting the *Times* on the back for calling

Trump a liar. "A precedent has been set," declared *CJR* approvingly. "Which is great."

CJR goes on to say, "By using a word that comes from the vocabulary of advocacy in its own voice, the *Times* and other news organizations have taken the truth-telling standards of the news business to a new level. Until the rise of Trump, only on rare occasions—criminal convictions and instances of plagiarism and falsifying résumés—were words like 'lie' ever used by the media in describing events in the news."

I am taken aback to see a journalism review publication explicitly applaud a news organization for using "vocabulary of advocacy" in its news reporting. These are novel times, indeed, for journalism.

Author and former *Times* reporter David Cay Johnston, a Pulitzer Prize winner, is quoted as saying he was "thrilled and flabbergasted" to see the *Times* call Trump a liar. Why would that be a "thrill"—unless one is steeped in bias and seeking to advance a narrative rather than report the facts?

In November 2018, Daniel Dale, a Washington reporter for the *Toronto Star*, also took on the role of anti-Trump activism, urging his media colleagues to join the party and invoke the "L" word against the president. "I think if we want to regain the trust that has been lost in media, we have to level with readers. We have to be seen to be straight shooters, and I think in those cases the word is lie," he said.

Talk about missing the point! In my view, the traditional media throwing around slanted accusations of "lies" at every turn only serves to further erode public trust in the media. In fact, Dale chipped away at his own credibility as an objective journalist by expressing abject bias against Trump, one of the political figures he covered. No matter; he was soon rewarded with a job at CNN, where his bio proudly states, "he was the first journalist to fact-check every false statement from Trump."

I checked Dale's Twitter feed on January 21, 2020, to see where his reporting focus lies now that he's at CNN. His bio shows that his

main focus is still fact-checking President Trump ("and other politicians"). A sample of his tweets includes:

> And when people say "people who think Trump is a liar already think he's
> a liar, so what's the point"—even people who already think he's a liar of-
> ten want to understand how precisely he is trying to deceive them on a
> particular subject.

> In mid-November, I published a list of 45 ways Trump had been dishonest
> about Ukraine and impeachment. It's now 65 ways. Here's the updated
> version, with quick fact checks of all 65 . . .

> Trump repeated one of his favorite tears lies this afternoon, the one
> about the people behind him crying as he signed his Waters of the United
> States order. (Unlike most of his tears stories, which tend to take place
> "backstage" and such, this event was on camera. Nobody cried.)

> Just some subjects of Trump dishonesty last week: Iran. Cancer. Ethi-
> opia. AOC. Bolton. NATO. Highway permits. Canada. His crowds. His ap-
> proval. His golfing. 2016. Vets. South Korea. Iraq. The visa lottery. China.
> Air. Obama.

Taking a closer look at Dale's Twitter feed, I see that Dale does
sometimes fact-check "other politicians," as advertised in his bio.
Of course, it often is little more than an academic exercise for
the purpose of trying to show, in yet another way, that Trump is a
liar. Here's what Dale concluded when he fact-checked the Democrats' impeachment leader, Congressman Adam Schiff, in January
2020:

> Schiff correctly pointed out that the (48) Republican members of the
> three committees holding the hearings were allowed in. (Non-members
> stunt-stormed a hearing in October and eventually left.) Schiff added,
> "And more than that: they got the same (questioning) time we did."

Schiff corrects [Trump White House counsel Pat] Cipollone's false claim that Republicans weren't allowed into the closed-door impeachment hearings. He says, "I'm not gonna suggest to you that Mr Cipollone would deliberately make a false statement . . . but I will tell you this: he's mistaken. He's mistaken."

In fact, Schiff himself has made numerous misrepresentations and false claims. For example, he mischaracterized a text message exchange and misidentified one of the officials mentioned in it. In doing so, he falsely implicated Trump's lawyer Rudy Giuliani. But news coverage of Schiff's errors received far less publicity than any mistake by Trump—real or imagined. Schiff also told conflicting and contradictory stories about his relationship with the supposed "whistleblower" who started the impeachment inquiry against Trump. And various Democratic officials also made misrepresentations when they repeatedly claimed during Trump's impeachment that certain facts were not in dispute or there was "no argument" when, in fact, the opposite was true. None of that interested the fact-checkers who were obsessively focused on Trump's "lies."

Meantime, in a bold and brave act of investigative fact-checking, Dale catches Trump lawyer Jay Sekulow in the most dastardly of misrepresentations: miscounting a number of days. Dale proudly tweets:

Sekulow said earlier, "33 days--33 days, they held onto those impeachment articles--33 days. It was such a rush of national security that -- impeach this president before Christmas that they then held them for 33 days."

It was 28 days.

He goes on to explain why he feels exposing this five-day miscalculation is so important:

All facts matter. Since Trump and company are consistently wrong in little ways, it's our job to point out how they're wrong in those little ways.

Think of it. Across the nation, news-reporting positions today are filled by supposedly talented journalists assigned the singular task of serving up the narrative of Trump-as-a-liar over and over again, rather than reporting news stories. Saving America and journalism one fact-check at a time!

"Lies" Redefined

We've established how the media quickly went from rarely calling a newsmaker a "liar" to doing it with such regularity that it rolls off their collective tongues seemingly without thought. With hearty support from journalism groups and their peers, it is understandable that reporters would grow even bolder. They began to expand the "lie" label to cases when targeted newsmakers are exaggerating, joking, or misspeaking, or when there is no possible way to know for sure whether a statement is true or false.

One example can be found in news reports declaring Trump was "lying" when he stated that many illegal immigrants had voted in the 2016 election. Certainly, one could argue the reporters had not seen evidence quantifying the problem. But they could not accurately state they knew Trump was lying when he made the claim.

Some of these stories calling Trump a liar relied on the fallacy that because only a limited number of fraudulent voters have been caught, it is a lie to state that "many" illegal immigrants voted. That reasoning is akin to claiming that the only people who speed are the ones who actually get stopped and ticketed. In fact, it is factually possible that many, if not most, illegal voters do not get caught.

Time magazine went so far as to post an Internet headline declaring "Trump Is Wrong—Non-citizens Don't Vote." The headline is provably false. There are many documented cases of non-citizens caught voting. For example, the *New York Times* reported on nineteen non-citizens caught voting illegally in North Carolina. In Texas, a state investigation found that 95,000 people described as non–U.S. citizens were registered to vote and about 58,000 of them voted in Texas elections between 1996 and 2018.

Even the body of the *Time* article contradicts its own headline that "non-citizens don't vote." It acknowledges that non-citizen voting does happen but claims it is rare. Again, none of the reporters has firsthand knowledge; they are deriving their conclusions from other people's reports and opinions. Yet they present their findings as if they were uncontested, confirmed facts. Whether you like Trump or not, this is poor journalistic practice and further erodes our credibility as an industry.

The media's eagle-eyed fact-checkers—sarcasm intended—were out in full force for Trump's 2019 State of the Union address. When the president claimed, "Our brave troops have now been fighting in the Middle East for almost nineteen years," they didn't let him get away with that whopper! You see, the actual length of time the United States had been fighting in the Mideast, they countered, was a bit more than seventeen years, not almost nineteen. *What a liar that Trump is!*

Actually, the media's own fact-check about the length of the Afghan War could be characterized as a "lie" by its own standards. The war began in October 2001. The State of the Union address in question took place more than eighteen years later—closer to Trump's "almost nineteen years" than the media's "seventeen years." But that does not fit the preferred Trump-is-a-liar narrative.

The crack fact-checkers also corrected the record when Trump referred to US forces fighting in the "Middle East" during his State of the Union address. Maybe Iraq is in the Middle East, the media agreed, but *Afghanistan is in central and south Asia! More lies from Trump! He's trying to fool us into thinking that Afghanistan is in the Middle East! But we're too smart for that!*

Actually, many people do consider Afghanistan to be part of the Mideast. The fact-checkers failed to note that Afghanistan was considered part of the Middle East before World War I and that the Bush administration had grouped Afghanistan into what it called the "Greater Middle East."

Then there was Trump's State of the Union claim that "Nearly 5 million Americans have been lifted off food stamps" over two years. Fact-checkers refuted the statement, arguing that the number of people on food stamps had decreased from 44.2 million in 2016 to 40.3 million in 2018. *That's a decline of only about 4 million, not the 5 million that Trump-the-liar claimed! Who does Trump think he's kidding?*

Actually, I checked and found government figures stating there were 39.7 million people on food stamps on average in 2018, down from 44.2 million in 2016. That makes Trump's figure spot on—not a "lie" at all—and it means that the fact-checking media were mistaken. But who fact-checks the fact-checkers?

And what's a good narrative if it doesn't bleed over into popular culture and entertainment? The left-wing *Esquire* has proved that non-news publications are happy to do their part for the cause. In January 2019, *Esquire*'s Dan Sinker pens a blog titled "In 2019, The Media Has to Do Better in Calling Out Trump's Shit. Being objective doesn't mean letting liars lie." In a stream-of-consciousness rambling, Sinker writes:

> *Big news organizations need to do a better job treating the President like the liar that he is.*
>
> *That he's a liar isn't a revelation, I know. News organizations have done an amazing job at tracking his lies (the* Washington Post *clocks him at 7,645 (!!!) since he took office, though it's probably higher since this was published), at fact checking his lies (Politifact ranks only 5 percent of his statements as true), at debating whether they should call them lies ("intent is key" decided NPR), and at inventing ranking systems to describe the volume of his lies (meet the "bottomless Pinnochio"). That's all good stuff.*
>
> *. . . when the holder of this particular office sends a series of unhinged tweets spouting laughably untrue fuckery, or takes questions outside Marine One and just spouts nonsense, or goes on a half-truth*

ramblefest at a cabinet meeting and the initial headlines and tweets that get sent out more often than not just follow traditionally "Jackhole Says" structure? Well, everyone gets screwed over"

One notable exception to the media outlets casually throwing around the "L" word is National Public Radio (NPR). When it came to Trump's allegation about illegal immigrants voting, NPR did not call it a "lie" but said there is "no credible evidence of widespread voter fraud."

That brings up a second set of news phrases the media invented and deployed specifically against Trump: "no credible evidence," "no evidence," and "without evidence."

"Without Evidence"

As a news journalist, I do not recall uttering or hearing the phrase "without evidence" when reporting on anybody or anything over the course of about thirty-five years. Until we covered Trump.

To show how suddenly and starkly the terminology emerged as part of the lexicon of journalists, I conducted a Google search using the words "Trump without evidence." It returned 179 million results. I then did the same search for "Obama without evidence," and none—zero—of the top results involved the media calling out Obama for claims "without evidence." In fact, searching for "Obama without evidence" returned primarily stories about, you guessed it, *Trump* having no evidence for his claims.

At first glance, it might seem as though phrases such as "without evidence" and "no credible evidence" are more fact based and responsible than tossing around accusations of "lies." But they can be just as problematic.

First, who decides what "evidence" is "credible"?

Second, absence of evidence does not necessarily mean a claim is discredited. After all, there was "no evidence" that polio could be transmitted via water—until there was. There was "no evidence" that some cholesterol is good for your health—until there was.

Third, "without evidence" is an invented concept for the purpose of slanting reporting. Throughout time, few newsmakers presented "evidence" when making statements. It was never expected that each comment or speech would be accompanied by a set of footnotes and citations. Until Trump. Now "without evidence" is commonly invoked in a one-sided fashion, usually against Trump and his supporters and typically when the media want to call them into question or disparage them.

This trend has helped ensure the media's downward spiral when it comes to the public trust.

Here are a few examples along with my notes. Remember, these phrases did not exist within newsspeak in any meaningful way prior to the press creating them to use against Trump. But you can see that once the word went out, everyone seemed to fall in line and pick up the jargon as if fulfilling orders from a Grand Poobah of Propaganda.

REUTERS: "Trump, Without Offering Evidence, Accuses Mueller of Crimes"
NOTE: It's worth noting that the media, "without evidence," widely accused President Trump and his associates of crimes, including colluding with Russia, being Putin stooges, and taking orders from President Putin.
REUTERS: "Trump, Without Evidence, Says Arizona 'Bracing' for Surge of Immigrants"
NOTE: The Reuters article was published in December 2018, when Arizona was indeed bracing for a surge of illegal immigrants, which shortly happened and was widely reported.
ABC7 NEW YORK: "Trump Claims, Without Evidence, That Mexico Will Pay for Border Wall Via Trade Deal"
NOTE: Trump's claim that a trade deal would result in Mexico paying for

a border wall is in the same category as President Obama making a forward-looking policy commitment such as "If you like your health insurance plan, you can keep it." However, only Trump's statements are reported as being "without evidence."

THE DAILY BEAST: "With Absolutely No Evidence, Trump Suggests U.K. Spied on Him for Obama"

NOTE: The above article referenced President Trump quoting a former CIA analyst. The Daily Beast attempted to disparage the idea by attaching the phrase "with absolutely no evidence" but did not apply the same standard to those—including some in the media—who made what turned out to be baseless accusations against Trump, such as that he had removed the bust of Martin Luther King, Jr., from the Oval Office or had not paid income taxes.

And there's one from the *Washington Post* that I find particularly remarkable: "Trump, Without Evidence, Claims His Campaign's Polling Shows Him Ahead in Every State Surveyed." The "evidence" is actually cited in the headline: Trump's campaign's polling. Whether one believes the evidence exists or not, a fairer headline would read, "Trump Claims Campaign's Unreleased Polling Shows Him Ahead in Every State Surveyed."

SUBSTITUTION GAME: To show how the media do not apply the same standard to their favored politicians, consider that Congressman Adam Schiff, a Democrat from California, repeatedly claimed for nearly two years (without evidence) that there was "ample evidence of [Trump-Russia] collusion in plain sight." The *Washington Post* did not require or provide evidence from Hillary Clinton when writing the headline "Hillary Clinton Accuses Trump of Being 'a Puppet' for Vladimir Putin." CNN, *U.S. News & World Report*, and the *Los Angeles Times* were apparently copacetic with House speaker Nancy Pelosi claiming, without evidence, that "Trump is 'engaged in a cover-up.'"

Trump remains the sole recipient of the "without evidence" treatment—all because of The Narrative.

One of the best examples is the police shooting of an eighteen-year-old unarmed suspect named Michael Brown on August 9, 2014, in Ferguson, Missouri. Brown was black. The officer who shot him, Darren Wilson, is white. The story of what really happened constitutes one of the most severe cases of media malpractice in our time. Uncorroborated witness accounts claimed Brown was raising his hands in surrender when he was shot. Those false reports were spread on the media worldwide, mostly unchallenged.

If ever there were a time for responsible journalists to present counterpoints or flag the inflammatory claims as unsubstantiated, this was the moment. But they didn't. The false claims against Officer Wilson were widely given uncritical credence. They prompted violent riots. They sparked an entire movement called "Hands up, don't shoot!"

Later, Obama's own Justice Department confirmed that the eyewitnesses implicating Officer Wilson were "unreliable." Their statements conflicted with one another and with previous statements they'd made. Reliable witnesses and forensic evidence supported Officer Wilson's account. Brown had reached into the police vehicle and grabbed Officer Wilson by the neck. Later in the confrontation, Brown had appeared to be lunging toward Officer Wilson when Officer Wilson shot him in self-defense. Those were the findings of the final Justice Department report issued in 2015. But the conclusions received nowhere near the publicity of the original false claims. There were no apologies to Officer Wilson. His career and life were ruined by false, slanted, irresponsible reporting.

To this day, many people still believe the Ferguson misconceptions. And the false narrative—that an innocent black youth was gunned down in cold blood by a white, racist police officer—continues to be disseminated in an organized fashion. On August 9 and 10, 2019, Democrats Elizabeth Warren, Tim Ryan, Cory Booker, Kamala Harris, Beto O'Rourke, Kirsten Gillibrand, Bernie Sanders, and

Bill de Blasio, all of whom were running for president at the time, tweeted statements containing disproven claims or implications, as if on cue.

ELIZABETH WARREN: 5 years ago Michael Brown was murdered by a white police officer in Ferguson, Missouri. Michael was unarmed yet he was shot 6 times. I stand with activists and organizers who continue the fight for justice for Michael. We must confront systemic racism and police violence head on.

CORY BOOKER: 5 years ago, Michael Brown was killed by a police officer. . . . I have been thinking all day about Mike and his family, and my prayers are with them. . . . I am also thinking about the everyday citizens who stood against this police violence and racism and were tear gassed for their patriotic acts. Ferguson called to the conscience of our nation and inspired a movement that rightly continues.

KAMALA HARRIS: Michael Brown's murder forever changed Ferguson and America. His tragic death sparked a desperately needed conversation and a nationwide movement. We must fight for stronger accountability and racial equity in our justice system.

BERNIE SANDERS: Michael Brown should be alive today. Five years after his death, we must finally end police violence against people of color.

KIRSTEN GILLIBRAND: 5 years ago, a Ferguson police officer killed Michael Brown, an unarmed teenager.
He shot him 6 times.
Nothing will bring Michael back, but we can't stop fighting the injustice done to his family and so many others.

TIM RYAN: Five years since the tragic death of Michael Brown and we still have significant work to do. We must rebuild trust between police and the communities they have sworn to protect.

BILL DE BLASIO: Michael Brown should be here today. My city knows the pain of Ferguson all too well.... NO ONE should die due to the color of their skin.

BETO O'ROURKE: Five years ago, Michael Brown was shot dead by a police officer.... we are reminded of an idea as urgent, and as ignored, today as it was when Michael was killed: Black Lives Matter.

There were no high-profile fact-checks by members of the national news media pointing out the outright falsehoods in these tweets. The claims were not only made "without evidence," they were contrary to the evidence. But the media were too busy fact-checking President Trump.

Likewise, when it comes to Trump-Russia collusion and other anti-Trump narratives, we in the media allowed two years of outlandish allegations to be made virtually free of challenge and without credible evidence. As with Ferguson, most of those claims proved to be untrue. "Trump may not have paid any income taxes for decades—if ever." "He never really wanted to be president." "He's lying about immigration being a problem." "He was Russian president Vladimir Putin's stooge. A spy for Russia." The list goes on. At the same time that we are calling out Trump, we are violating the very evidentiary standards to which we hold him. In our zeal to get Trump, we let our own journalistic principles slide. We have laid bare our own bias and double standards.

We can blame Trump all we like for the death of the news as we once knew it, but the truth is: we've done it to ourselves.

Lies vs. Gaffes

Donald Trump twice claimed that, as president, he met with students who survived the mass shooting in Parkland, Florida. But it turns out he wasn't even president when the shooting occurred!

Okay, that didn't happen. Trump never made such claims. I substituted Trump's name for the actual fact-challenged offender: Joe Biden.

In reality, it was Joe Biden who twice claimed that, as vice president, he'd met with survivors of the mass shooting in Parkland, Florida. "I met with [the Parkland kids] and then they went off up on the Hill when I was vice president," he told the audience at a gun control forum in Iowa in August 2018. Later, he said, "Those kids in Parkland came up to see me when I was vice president." But Biden was inexplicably confused; he had actually been out of office a year when the Parkland shooting happened, so he could not have possibly met with the survivors when he was vice president.

The reason I put Trump's name into Biden's place—a classic Substitution Game exercise—is so that you can imagine for yourself how differently the incident would have been treated by the national press if Trump had made the mistake rather than Biden. But because it was Biden, the media collectively decided—all using the exact same word—that it was simply a "gaffe." A harmless, unintentional error. *Not like all of those willful, malicious lies that Trump tells.*

The widespread use of the word *gaffe* is itself evidence of the presence of a narrative. Here is how I know: It is not a word people commonly use in everyday conversation. We might ordinarily use words and phrases such as "error," "slip of the tongue," "mistake," or "mess up." Most folks don't say, "I made a horrible *gaffe* when talking to my uncle at work." So why is it the accepted word that nearly all the press adopts when discussing Biden's blunders?

Interestingly, "gaffe" is the very word Biden uses to refer to his own penchant for saying things that turn out to be either offensive or incorrect. Early in Campaign 2020, he misstated how many times he'd been to Afghanistan and Iraq, exaggerating the number by one-third (claiming he had visited "over thirty times" instead of twenty-one). He mixed up British prime ministers. He confused Charlotte, North Carolina, with Charlottesville, Virginia. He greatly overstated the number of people shot at Kent State in 1970 (he said it was more than fifty, but it was actually four). He later mistakenly told a crowd he was running for US Senate (instead of president of the United States), claimed to have negotiated with a Chinese dictator who'd been dead for decades, and stated that half the country, "150 million people," had been "killed by gun violence since 2007." On March 2, 2020, he tried to quote the Declaration of Independence, but it came out as "We hold these truths to be self-evident . . . all men and women are created . . . by . . . go . . . you know, the, you know the thing." At campaign appearances in March 2020, he mixed up his sister and his wife, and referred to AR-15 rifles as "AR-14s."

Even when Biden makes so many "gaffes" in a single story that the news media tally them up, they still do not characterize the errors as anything sinister or intentional—not that they are, but this is in stark contrast to the slanted treatment Trump gets. On August 29, 2019, the *Washington Post* notes that Biden managed to make a half dozen factual errors in one short war story. It happened at a New Hampshire town hall meeting when Biden recounted an emotional tale about a veteran and apparently got wrong the time period, medal, rank, military branch, location, and heroic act involved. CNN gently characterized the mistakes as "several inaccurate elements," "misstatements," and "misrememberings" and said that Biden had been "incorrect."

As if in a parody of itself, the discredited fact-checker Snopes actually said that the *Washington Post* and others were in error to imply that Biden's war story was false because *parts of his story were true*. Snopes twists itself into a pretzel to please its narrative masters: Biden is to be legitimized, not criticized.

So although it is a legitimate pursuit for the press to examine and call out false statements by political figures, we destroy our own credibility by not treating similar false statements equally. Contrast Biden's "gaffes" with what is said about Trump. Biden is a "gaffe machine"; Trump is a "congenital liar." Take the *New York Times*. In a list it compiled in 2017 of "Trump's Lies," it stretched the definition of "lie" by using the term to describe statements that are obvious mistakes or exaggerations.

Here is my brief analysis of seven "lies" as compiled in the *New York Times'* list of "Trump's Lies."

1. TRUMP LIE: "Between 3 million and 5 million illegal votes caused me to lose the popular vote."

   NEW YORK TIMES: "There's no evidence of illegal voting."

   ANALYSIS: Absence of evidence is neither proof it did not happen nor proof of a lie. The *Times* fails to acknowledge evidence, such as an academic study that previously found millions of illegal votes "likely changed 2008 outcomes including Electoral College votes and the composition of Congress" in favor of Democrats. Further, the *Times* presents "no evidence" for its own claim debunking Trump. That means—applying the *Times'* own standard—that the newspaper is "lying" about Trump.

2. TRUMP LIE: "ICE [Immigration and Customs Enforcement] came and endorsed me."

   NEW YORK TIMES: "Only its union did."

   ANALYSIS: Since the union for ICE endorsed Trump, it's hardly a "lie" for him to state that ICE endorsed him. At worst, one could fairly say it is an overstatement or exaggeration.

3. TRUMP LIE: "With just one negotiation on one set of airplanes, I saved the taxpayers of our country over $700 million."

   NEW YORK TIMES: "Much of the cost cuts were planned before Trump."

   ANALYSIS: The *Times* could have stated that it believed Trump was claiming some credit for cuts planned by someone else or exaggerating the value of his negotiations, but his statement does not qualify as a "lie."

4. TRUMP LIE: "Now, my last tweet—you know, the one that you are talking about, perhaps—was the one about being, in quotes, wiretapped, meaning surveilled. Guess what? It is turning out to be true."

   NEW YORK TIMES: "There is still no evidence."

ANALYSIS: The *Times* may disagree with what constitutes "evidence," but there is significant evidence that Trump and his campaign were surveilled and wiretapped in multiple ways. His statement cannot fairly be termed a "lie."

5. TRUMP LIE: "We are 5 and 0 . . . in these special elections."

NEW YORK TIMES: "Republicans have won four special elections this year, while a Democrat won one."

ANALYSIS: Trump's mistake was saying "five" instead of "four." Unless the *Times* has proof that Trump set out to deceive and knew that the correct number was four rather than five, his statement is most accurately described as an error or misstatement—not a "lie."

6. TRUMP LIE: "I mean truly dishonest people in the media and the fake media, they make up stories. They have no sources in many cases. They say 'a source says'—there is no such thing."

NEW YORK TIMES: "The media does not make up sources."

ANALYSIS: The *New York Times* is just plain wrong. First, it cannot possibly claim to know about and speak for every member of the media. Further, there are multiple known instances of the media making up sources. High-profile examples include Janet Cooke, whose Pulitzer Prize (won while she was at the *Washington Post*) was revoked when it was discovered she had fabricated her main character in a news series; the *New York Times*' own Jayson Blair, who turned out to be a serial fabricator and plagiarist; the *Boston Globe*'s Mike Barnicle, who made up sources and facts in a story about children with cancer; and CNN International's "Journalist of the Year," Claas Relotius of *Der Spiegel*, who got caught fabricating a dozen anti-Trump stories and people in them.

7. TRUMP LIE: "The Russia story is a total fabrication."

NEW YORK TIMES: "It's not."

ANALYSIS: The *Times* can express an opinion that the Russia collusion story was not a fabrication or was not a "total" fabrication, but Trump's opinion cannot objectively be called a "lie."

CNN, likewise, does not bother to hide its double standards. In one article, it explicitly argues that voters should make a distinction between "Biden's slips," and Trump's "cavalcade of lies [and] purposeful daily assaults on the truth." But an examination by RealClearPolitics proves the folly of CNN's distinction. RealClearPolitics notes that news headlines labeled "Trump's errors as 'bizarre,' 'baffling,' 'bungles,' and 'whoppers' whereas his predecessor Barack Obama's errors were termed as merely 'misleading' or 'cherry-picked.'" Yet RealClearPolitics' analysis of fact-checks over a four-month period found that Trump's statements were not significantly different from Biden's. Both Biden and Trump had 38 percent of their claims labeled "False."

The difference in the way misstatements are reported in the news depending on who makes them can be explained only through The Narrative. Trump is a devious, evil devil; Biden is a well-meaning, affable, innocent goof.

Of course, Trump has one big advantage that enables him to combat the negative treatment he gets at the hands of many reporters and vex them at the same time: he is a narrative machine.

The Narrative Machine

Many propagandists work in a clandestine manner, cleverly inserting narratives into our consciousness so that we believe them without knowing where they started or why. It is a critical part of their tradecraft to make sure their product is noticed while their own hand in it remains hidden.

But it is different with President Trump. He is the first public figure to have incredible success using narratives in such an upfront fashion. His efforts are very much "in your face." He leaves no doubt that he is intentionally working to plant narratives. He uses his tradecraft against Democrats and Republicans alike, particularly if he feels they have attacked him first.

As much criticism as the practice has drawn, it has generated far more in the way of dividends for Trump. The fact that so many people can rattle off many of the narratives he drives is one indication of their success.

Trump applies his narratives through labels that are succinct, catchy, and easy to understand. He drives them home with endless repetition. Sometimes he incorporates humor into them, making them all the more memorable to his fans.

"POCAHONTAS": The name assigned to Elizabeth Warren, who ran for the Democratic nomination for president and had falsely claimed to be of Native American heritage, drove home the narrative that she is dishonest and a hypocritical cultural appropriator.

"SLEEPY JOE": Former vice president Joe Biden was tagged as too weak and unenergetic to be commander in chief.

"LOW ENERGY JEB": Once Trump made this label stick to Republican presidential candidate Jeb Bush in 2016, it was hard not to watch Bush's understated presentations and not think about it.

"CRAZY BERNIE": Trump dismissed Senator Bernie Sanders, running for president for a second time in 2020, by conveying the narrative that his socialist-leaning ideas were unhinged.

"CRAZY NANCY": The president also hit House speaker Nancy Pelosi with the "crazy" moniker.

"CROOKED HILLARY": Through these two simple words, Trump was able to evoke all of the alleged crimes, conspiracies, and wrongdoing attributed to former first lady Hillary Clinton.

"CRYIN' CHUCK": In 2017, when the Senate's lead Democrat teared up over Trump's "mean-spirited" immigration ban, he gave Trump a gift that keeps on giving. Sometimes Trump also uses "Lyin' Chuck."

Even when Trump uses phrases that are degrading physical descriptions, they say it all. And more important: they stick. He sometimes uses the imagery of physical smallness to undermine the perception of his critics' and opponents' strength and authority: "Little Marco" for his onetime Republican opponent in 2016 Senator Marco Rubio; "Mini Mike" for Democrat Michael Bloomberg, running against Trump in 2020; and "Pencil Neck" Adam Schiff for the House Democrat who led impeachment efforts against Trump.

Trump's "Fake News" and "Enemy of the People" narratives against the press have become particularly ubiquitous, playing off preexisting public skepticism of the news media. As I note in my previous book, *The Smear*, the modern use of the phrase "fake news" was not Trump's invention. The effort to define and crack down on "fake news" was launched during the 2016 campaign by the nonprofit website First Draft, which was funded by Google, which is owned by Alphabet (a top supporter of Hillary Clinton and Bernie Sanders), which was led at the time by top Hillary donor Eric Schmidt. Under their definition, fake news was always conservative in nature. Shortly after First Draft began pushing the "fake news" narrative, President Obama drove it home in an October 13, 2016, speech at Carnegie Mellon University. As I mentioned earlier, Obama further claimed that somebody needed to step in and "curate" information online for the public's own good. This was the beginning of an effort to convince us to accept third parties, whether government, corporations, academics, or social media companies, deciding what information we should have and telling us what we should believe.

It did not take long for Trump to prove he is better at the game. He co-opted and redefined "fake news," turning it against its creators, who now disavow it. These days, ask most people, and they mistakenly think Trump invented the phrase. Actually, it was just a hostile takeover demonstrating his mastery of The Narrative.

Equally as impactful are other catchphrases Trump has invented. To black Americans whose votes Trump courted: "What have you got to lose?" Referring to "Crooked" Hillary Clinton at his rallies: "Lock her up!" To hecklers? "Go home to Mommy."

"Make America Great Again." "We're not winning anymore." "Witch hunt."

Trump supporters are delighted when he repeats stories and mantras. In 2016, he frequently told a version of Aesop's fable about an old woman who invited a snake into her home and should not have been surprised when it turned on her. At nearly every rally, he directs the crowd to look behind them at the "fake news" cameras as he chides their operators for refusing to show how big his crowds are.

With Trump's political opponents and much of the news media looking unfavorably upon the narratives he devises and the way he executes them, how has he been so successful at using them?

Obama once said that to forward his agenda, he had "a phone and a pen." That referred to his ability to call people to build support and to veto measures he didn't like. But all of that is so yesteryear. Trump has something that is arguably even more effective: Twitter. Twitter's space constraints provide the perfect vehicle for a man who distributes narratives in short phrases and bursts. With more than 72 million followers on his @RealDonaldTrump account and 28 million more on his White House account, that's a neat 100 million follows. Trump is able to speak directly to both friend and foe in an instant to drive home his many narratives.

A second tool at Trump's disposal is the traditional bully pulpit. Whether they like it or not, the media have to cover many of his events. It could be an event to honor sports figures or war heroes, a meeting with a foreign dignitary, or an impromptu stop at the microphones on his way to catch a flight to Air Force One—Trump rarely misses an opportunity to get in a dig or reinforce a narrative. He knows there will be many eyeballs focused on him and ears listening. He treats his bully pulpit as if it were an episode of a reality show. When asked what he might do or how something might turn out, he often answers with phrases such as "We'll see" or "You'll see pretty soon." *Tune in next time.*

During about six weeks of the coronavirus crisis in 2020, from March 13 to April 23, I added up a little more than forty-one and a half hours President Trump spent talking to the public on camera and taking questions from reporters. That averages to about an hour a day, seven days a week. I can't think of any other president in our time who's done anything close to that. He often spent time driving home the same set of talking points. On the anti-malaria medicine hydroxychloroquine as a possible treatment or prevention: "It may work; it may not . . . we'll see . . . but what have you got to lose?" On the widely reported fears of a ventilator shortage that he managed to help backstop: "The press won't give me credit . . . nobody who needed a ventilator was denied one."

In an interview I conducted with President Trump at the White House in May 2020, I asked him why he chose to spend so much time in front of TV cameras and if he felt it was time well spent.

"Well, I think it was," he told me. "I certainly got the highest ratings on cable television by a lot. I mean, you saw that." He also pointed out that it was his way to "get to the public" around the "very corrupt" news.

Certainly, Trump's enemies have also enjoyed some success using repetition and catchy phrases to tarnish the president with various narratives. Early on, starting in 2015, when Trump began to first look like a serious political challenger, both Democrats and Republicans tried out a series of narratives. In the beginning, Trump was frequently described as a "clown," advancing the narrative that he was not to be taken seriously. When he became a real contender, the phrasing quickly switched to "dark and dangerous," furthering the notion that he could not be trusted to be president.

But four narratives promulgated against President Trump in an organized fashion turned out to be more successful than the others.

First, not long before the election in 2016, the liberal smear group Media Matters and its affiliates started a "white nationalist" narrative against Trump and his associates that morphed into "white supremacist" and "racist" labels. Prior to that time, there was virtually no public accusation of this kind to be found against Trump; in fact, quite the opposite: he was frequently celebrated by notable black media and political figures and portrayed favorably in the popular press.

Second, Trump's opponents managed to twist Trump's pro–legal immigrant, anti–illegal immigrant stance into a broad position, as if he were "anti-immigrant" and "racist." I always considered this particularly contrary to facts, since nearly every time Trump speaks of immigration, he talks about how he values legal immigration. Not to mention the fact that he married two immigrants and his children are the children of immigrants.

Third, as we have already discussed, not a day goes by without media and political figures calling Trump a "liar."

And fourth, as I have briefly mentioned, there was the narrative of Trump as a "Russian stooge." Even after Special Counsel Robert Mueller concluded there was no evidence that Trump, his campaign, or any American "colluded with Russia," the label still sticks among Trump opponents. I call it "The Mother of All Narratives."

CHAPTER 7

The Mother of All Narratives

RUSSIA, RUSSIA, RUSSIA

In a book about declining media credibility, the biggest narrative of all deserves special examination. It is the story line that Donald Trump somehow colluded with Russian president Vladimir Putin in order to win the US presidency in 2016. Trump-Russia collusion will go down in the history books (if they were to write of such things) as one of the most successful narratives in modern times in terms of its dominance, its pervasiveness, and its treatment as breaking, front-page news. It is a stunning feat of propaganda. A Google search for "Trump Russia collusion" in the second half of 2019 returned 10.5 million hits in less than half a second. I think this particular narrative is responsible for the single greatest erosion of public trust in reporting by mainstream news organizations.

For more than two years, reporters and pundits insisted Trump had conspired with Russia to win the presidency, even though there was no publicly available proof of any such thing. It is unprecedented how formerly well respected national news organizations justified suspending long-standing ethics and journalism guidelines in order to promote a slanted and ultimately false story line. The Narrative, perpetuated by the media and US intelligence officials, took on an incredible life of its own. The more false it became, the more undeniable it seemingly grew. At any point, it would have been easy for the media to step back, follow professional practices, and put the facts into context and perspective. Instead, we risked our very jobs and credibility in our zeal to sell the public a bill of goods.

You know how it ended. Even with Trump's political enemies assigned to the team investigating him, even with the Department of Justice inspector general finding egregious abuses committed by the FBI and Justice Department investigating the Trump campaign, even with an FBI lawyer admitting to doctoring a document to justify an improper wiretap on a former Trump campaign volunteer, Special Counsel Robert Mueller was unable to produce evidence of Trump colluding with Russia. But that conclusion wasn't met by the press with embarrassment over their role, apologies for their mistakes, or even a fleeting expression of regret. They quickly moved on to the next narrative.

I think the biggest victim of the whole Trump-Russia narrative—even bigger than Trump himself, who somehow managed to survive relatively unscathed, all things considered—is a man named Carter Page. If we hadn't been blinded by the prevailing narrative, Page's sordid tale would be one of the biggest stories of the 2016 election. Since you didn't hear much about it on the news, it is worth mentioning now.

If you did hear anything about Carter Page in the popular news media, you likely think he's either a moron and a buffoon with a low IQ or a Secret Russian Agent Man shrewd enough to mastermind an international plot to put a Manchurian candidate into the White House.

I found Page unlikely to be either of these things. I had the chance to meet him myself when I set up an interview with him in March 2019 for my Sunday news program, *Full Measure*. It was shortly before the Mueller Report was released, and I saw the writing on the wall: Page had been smeared by slanted news coverage and The Narrative. His reputation was ruined. But after he was hounded by the press and subjected to the government's most invasive kind of secret-squirrel tactics for months upon years, I could see that the government was not about to find anything to charge him with.

I first meet Page for an interview at my studio offices in Arlington, Virginia, just a few miles from the nation's capital. He is friendly and offers a smile. He's wearing a custom-made dark blue suit, a crisp white shirt, and a midnight blue tie with thin diagonal white stripes. Page is slender, neat, and polite, shaves his head bald, and is about to turn forty-eight. He proves to be soft-spoken, restrained, and thoughtful. Rarely reported among the media narratives is the fact that he served as a naval officer in Europe and the Mideast with a brief stint in Navy intelligence. He also earned two master's degrees and a PhD, became a successful investment adviser, and worked as a businessman in Russia from 2004 to 2007.

How silly, Page comments in the interview, that the media now assigns something nefarious to his business ties to Russia. He worked as an executive with Merrill Lynch, assigned to the company's Moscow offices. US business relationships with Russia are strikingly common. Contrary to all the conspiracy theory narratives, the US government actually encouraged business relationships with Russia after the fall of the Soviet Union. Our government even created opportunities for US businesses to get involved with Russia to help integrate it into the Western economy. Page remarks that at the time he was developing his Russian links, it was considered a patriotic thing to do. That was before The Narrative claiming his Russia ties proved he was a Russian spy.

During the course of my interview with Page, I learn two stunning facts I hadn't heard widely reported. First, Page has a long history of assisting US intelligence agencies, including—wait for it—*on Russian spy cases.* I discover a second, even more amazing fact when I ask Page when he first met Trump in person.

"I never met him at all," Page replies.

"*You never met Donald Trump?*" I ask. I'm not sure I'm hearing him correctly.

"No," says Page.

"Never spoke to him?"

"Never," Page confirms. "Never on the phone. Nothing." Not to this day.

How could the FBI's number one suspect as the supposed go-between for Putin and Trump be someone who never even met Trump?

As I continue the interview, my thoughts are racing. I review the FBI's theory in my mind. It claims that a guy who didn't know Trump and who had helped the FBI and CIA in the past, including with Russian spy cases, himself became a Russian spy while knowing he was under active surveillance by the FBI?

"It's just so outrageous, preposterous. Where do you even begin?" Page asks.

Perhaps a good place to start is June 16, 2015, when Trump descended the escalator in Trump Tower in New York City to announce his candidacy. That's the day, Page tells me, he decided to volunteer for the Trump campaign: "I think President Trump, then candidate Trump . . . had a great vision for the direction that the world should head and the US role in it. And I wanted to help out in any way that I can." He goes on to say he eventually connected with other Trump campaign volunteer advisers on foreign policy issues. He had no idea he would soon be at the nexus of the Trump-Russia collusion narrative.

The abbreviated version of the whole sordid mess is that once Page got involved in the Trump campaign, he became targeted by an anti-Trump political opposition research effort funded by Democrats and the Clinton campaign. Their hired guns, a company called Fusion GPS, hired an ex–British spy named Christopher Steele. Steele collected rumors and dirt about Trump and Page from Russian operatives. Various foreign figures and members of Congress were used to put the file, the so-called Steele dossier, into the hands of the news media and the FBI to implicate Page and Trump in all kinds of Russian mischief and possible crimes.

Six weeks before the 2016 election, Yahoo! News published the

Steele dossier. In response, Page wrote a letter to then FBI director James Comey telling him "just how absolutely outrageous this whole thing is.

"I mentioned . . . the fact that I had helped CIA and FBI over many years, and I said, 'We've had long conversations with the intelligence community. . . . I mean, this is just so implausible on the face, but if you have any questions about it whatsoever, please do not hesitate to contact me.'"

The FBI ignored Page's outreach and doubled down. The following month, the government secretly obtained a Foreign Intelligence Surveillance Act (FISA) court warrant to spy on Page. Using the dossier as part of the evidence—in violation of strict FBI rules that prohibit presenting the court with even a single unverified fact—the FBI convinced the court that Page and perhaps others in the Trump campaign were "collaborating and conspiring with the Russian government."

It's worth noting that before our intelligence agencies take the drastic step of invading the constitutional privacy of US citizens by deploying government intel tools against them as they did to Page, the FBI is required to have independent evidence in hand that the target is acting as a foreign spy or is imminently about to become a foreign spy. Somehow the FBI got the FISA court to sign off on the false notion about Page time and time again, even though the FBI had no actual "goods" on him. There were four FBI wiretaps against Page for ninety days each. Much later, after an investigation by the inspector general, the wiretapping was ruled to have been improper and conducted by FBI officials who committed egregious violations.

What's even more important is the spin-off privacy violations committed by the government because of the false narrative about Page being a Russian spy. Under a little-known government policy at the time, wiretaps against one target (such as Page) allowed our intel agencies not only to collect emails, phone records, bank records, text messages, photographs, and other communications belonging to the target but also to rifle through the same personal material belonging to people as many as "two hops" away from the target. That means the court-approved wiretaps against Page could also have been used to collect highly personal information on anyone who communicated with him (one "hop"), and anyone who communicated with *that* person (two "hops")—even if the people two hops away had never communicated with Page!

You can start to understand how one analysis of this policy concluded that intel agencies could exploit one legal wiretap to secretly access 25,000 people's phones. It's yet another reason why wiretaps on any American citizen are supposed to be pursued judiciously, cautiously, and conservatively. But when the dust settled on the investigation into Trump, it sure looked a lot like anti-Trump intel officials had wiretapped Page to capture private communications of Page's contacts and *their* contacts, including Trump himself.

"You communicated with people, including [Trump adviser] Steve Bannon, who were talking to President Trump?" I ask Page.

"Yes," he replies.

"Therefore, Trump would have or could have been wrapped up in the same surveillance?"

"Absolutely. Absolutely," answers Page.

I don't think the wiretaps against Page were the only vehicle that rogue actors in our intelligence agencies exploited to try to get Trump in 2016. According to news reports, more than half a dozen people surrounding Trump were captured on FBI wiretaps during that time period. This is nothing short of astounding.

According to media reports in September 2017, the FBI also wiretapped the former head of Trump's campaign, Paul Manafort, both before and after Trump was elected. Intel officials captured former Trump adviser Lieutenant General Michael Flynn on electronic surveillance. Former Trump campaign adviser George Papadopoulos has reported that he believes he was surveilled. Multiple Trump "transition officials" were "incidentally" picked up during government surveillance of a foreign official. We know this because former Obama adviser Susan Rice reportedly admitted "unmasking," or asking to know the identities of, these officials. In May 2017, former director of national intelligence James Clapper and former acting attorney general Sally Yates acknowledged that they, too, had reviewed communications of certain unnamed political figures secretly collected under President Obama. Trump associate Roger Stone was also reportedly picked up on wiretaps.

Let's see: at least six Trump associates wiretapped, multiplied by 25,000 people possibly surveilled "two hops" away from each target, equals 150,000 people possibly spied upon by our government . . .

Back to Carter Page.

After all the unconfirmed rumors and allegations by anonymous sources surfaced in the press in September 2016, Page left his volunteer position with the Trump campaign. But the government secretly continued surveilling him—even after Trump was inaugurated. For one full year, Page's every move was watched: every call presumably listened in on; every move he made online subject to review by nameless, faceless government agents. After no charges against him came, the media had a narrative for that, too, other than his innocence. Here's how the speculation played out on CBS on July 23, 2018.

"If the FBI had reason to believe Carter Page was acting as an agent of Russia, why isn't he facing any charges?" a CBS anchor asks New York University law professor Ryan Goodman. It's a logical question. But it evokes a counterintuitive answer—counterintuitive, that is, but for The Narrative.

"You might . . . want to have someone who is a suspected criminal roaming free because if they're under surveillance you're picking up a lot of valuable information," reasons Professor Goodman.

I find that analysis absurd. By this time, the whole world, including Page himself, knew he was under surveillance. For Goodman's theory to be correct, we would have to believe that Page is actively spying for Russia while knowing he is under watch by the FBI. There is no mention in the CBS news report that Professor Goodman is a liberal Trump critic who worked for the Obama administration in the Pentagon's Office of General Counsel. Shouldn't viewers be provided the professor's background and interests—and maybe even be offered an analysis from the other side for balance? Of course. But that would undercut The Narrative. In a slanted news environment in which The Narrative is the goal, Professor Goodman's one-sided analysis makes perfect sense.

Other media, including CNN, join the club. "Despite what a 400-page document suggests, Carter Page says he is not an agent of Russia," reports CNN on July 18, 2018. The headline lead mentioning the "400-page document" casts doubt on Page's denial. Whatever supports The Narrative is automatically awarded credibility; anything that fights it is treated as questionable.

Despite all he's been subjected to, Page sounds calm and relatively upbeat when he speaks with me in the interview. "My biggest concern throughout this has been the damage that it's done to the country," he tells me. "And so I always sort of laughed off [the government's conspiracy theories about me]. And I think . . . that was a negative cycle in a way. Because, sometimes, if I'm laughing at these people, they almost want to come after you even harder to really bring you down. But I was always more concerned about the damage that it was doing to the Trump administration and other people."

Narrative Traffickers

When we can find out who, exactly, is pushing a given narrative, it tells us a lot about the truth of the matter. When it comes to the Mother of All Narratives, we are offered a unique glimpse behind the curtain because we have learned so much about the puppet masters. Their identities explain a great deal about how they were able to command the media stage, execute The Narrative with chilling precision, and become wildly successful at convincing news consumers to buy into it.

The tools at their disposal included access to government insiders, newspapers that were anxious to publish their op-eds, a press eager to report their anonymously leaked information (some of it true and some of it false), social media campaigns, and news appearances as analysts and commentators.

Much as Trump prompts his audience to "stay tuned" to the next episode of his presidential reality show, the traffickers of the Russia collusion narrative deployed a similar strategy. They appeared daily on cable TV news and at press conferences to make claims and give hints. They made cryptic references to as-yet-unrevealed secret information that they implied proved the crimes. *Stay tuned, there's more . . .*

Dozens of key players were involved in the effort. Each proved to be a crucial figure in both the development and deployment of the slanted Trump-Russia narrative. Each was front and center in both the creation and delivery phase of The Narrative. And as you'll see, each had more than a passing interest in selling the public a bill of goods. In some cases, they needed The Narrative to deflect from their own long-standing misdeeds prior to 2016—those that Trump and his then right-hand intel man, Lieutenant General Michael Flynn, threatened to expose. We know from private text messages that these players hoped The Narrative would prevent Trump from being elected. That would have cleared the deck for a different president who would not challenge the intelligence community's status quo or dig into the dark recesses to learn what dastardly deeds had been committed in the past. Once Trump foiled their plans and got elected anyway, they needed The Narrative to prevent discovery of their operations against him. Ultimately, the 2016 Trump-Russia narrative looks like an operation to cover up a cover-up of an operation, if you will.

But these players involved proved to be the gang that couldn't shoot straight. Trump got elected, was cleared of Russia collusion by Special Counsel Robert Mueller, wasn't ousted by impeachment, and their scheme was exposed. On the other hand, they proved to be experts at promulgating false narratives. And so, with the help of a complicit press, their antics were hidden and spun until, to this day, many in the public remain confused or ignorant about exactly what happened. The collateral damage is our faith in the very institutions designed to protect us.

Here are five of the highest-profile players among the Trump-Russia narrative traffickers.

James Clapper

KEY BACKGROUND: Director of National Intelligence (DNI) James Clapper, appointed by President Obama, falsely assured Congress in 2013 that the NSA was not collecting "any type of data at all on millions or hundreds of millions of Americans." NSA whistleblower Edward Snowden's revelations just weeks later proved Clapper's testimony to be false. (Clapper then apologized to Congress, saying he'd misunderstood the members' questions.) Under Clapper, US intel agencies secretly monitored conversations of members of Congress while the Obama administration negotiated the Iran nuclear deal. Clapper wasn't the first to be in charge while our intel agencies conducted questionable surveillance. In 2011, under President Obama, our intelligence agencies wiretapped Democratic congressman Dennis Kucinich of Ohio. In 2004, under President George W. Bush, Democratic congresswoman Jane Harman of California was surveilled. In both instances, someone leaked information captured on the secret wiretaps so that it ended up in the press. In any event, documents show that under Clapper's reign, intel agencies vastly expanded their encroachments on US citizens' privacy.

ROLE IN THE NARRATIVE: Clapper resigned from his leadership role in the intelligence community in 2016 after Trump was elected, and was hired as a CNN national security analyst. There he became an ever-present and vocal critic of President Trump, helping to execute the false Russia collusion narrative. As George Washington University law professor Jonathan Turley pointed out in an opinion piece for The Hill, "After leaving as DNI, Clapper was used repeatedly by CNN without mentioning his alleged perjury on the surveillance program. CNN, for example, did not mention it in using him to rebut Trump's allegation that his campaign staff was surveilled under the Obama administration; Clapper categorically denied it and said he would have been aware of such secret surveillance. In fact, Trump associates, including Carter Page, *were* under surveillance." Turley says that in another instance, "The report [by Special Counsel Mueller] recounts how Clapper gave 'inconsistent testimony' to Congress when he denied ever 'discuss[ing] the dossier or any other intelligence related to Russia hacking of the 2016 election with journalists.' That also has proven to be untrue." Turley goes on to say, "Clapper later admitted he discussed the 'dossier with CNN journalist Jake Tapper' and indicated he may have discussed the material with other journalists."

On NBC's *Meet the Press*, Clapper continued implicating Trump, telling viewers that his "dashboard warning light was clearly on," regarding possible contact between Russians and Trump officials. He also told reporters, "I think if you compare the two that Watergate pales, really, in my view, compared to what we're confronting now." On CNN in December 2017, he called Vladimir Putin a "great case officer . . . he knows how to handle an asset, and that's what he's doing with the president." In 2019, also on CNN, he said "it was a possibility" that Trump was a "Russian asset."

Later, we learned that Clapper had given closed-door testimony under oath to Congress in July 2017, admitting he "never saw any direct empirical evidence that the Trump campaign or someone in it was plotting [or] conspiring with the Russians to meddle with the election."

Rarely did reporters challenge Clapper's information and claims along the way. Never did the media modify his unsupported statements with the phrase they use against President Trump: "without evidence." And when Clapper's analyses and information proved wrong, there he was the next hour on CNN, relied upon for still more analysis.

John Brennan

KEY BACKGROUND: In 2014, CIA director John Brennan got caught red-handed spying on Senate Intelligence Committee staffers. Like Clapper, he explicitly denied the deed. However, when the inspector general confirmed it was true, Brennan issued an apology, and Congress seemed to forgive and forget the transgression. As was the case with Clapper, Congress's inaction regarding the false testimony sent an implicit message to the bad actors that they can "carry on," and Congress awarded them ever-expanding access to sensitive information.

Brennan was associated with questionable political operations against journalists going back at least to 2010, prior to his leadership at the CIA. At the time, he was a deputy national security advisor for homeland security and antiterrorism under President Obama. An internal email at the "global intelligence" firm Stratfor, dated September 21, 2010, and exposed by WikiLeaks, alleged, "Brennan is behind the witch hunts of investigative journalists learning information from inside the beltway sources. *Note—There is specific tasker from the WH to go after anyone printing materials negative to the Obama agenda (oh my.). Even the FBI is shocked. The Wonder Boys must be in meltdown mode . . .*"

Also, with Clapper and Brennan at the helm, there were shocking findings by the secretive Foreign Intelligence Surveillance Court (FISC) in 2016. A judge found egregious violations of strict surveillance procedures by the intelligence community during the election year.

ROLE IN THE NARRATIVE: Brennan resigned in January 2016. While Clapper manned the CNN airways, Brennan went over to play the same part at NBC News and MSNBC, where he was hired as a senior national security and intelligence analyst. He also became a tweeting machine the likes of which have never before been seen from a former head of an intelligence agency. Highlights of some of his tweets and attacks on President Trump include the following.

MARCH 2018: Brennan called Trump a "charlatan," said he suffered "paranoia," and accused him of "constant misrepresentation of the facts." That same month, he tweeted to Trump:

> When the full extent of your venality, moral turpitude, and political corruption becomes known, you will take your rightful place as a disgraced demagogue in the dustbin of history. You may scapegoat [FBI deputy director] Andy McCabe, but you will not destroy America . . . America will triumph over you.

JULY 16, 2018: Brennan tweeted:

> Donald Trump's press conference performance in Helsinki rises to & exceeds the threshold of "high crimes & misdemeanors." It was nothing short of treasonous. Not only were Trump's comments imbecilic, he is wholly in the pocket of Putin. Republican Patriots: Where are you???

DECEMBER 31, 2018: Brennan tweeted to Trump that he hoped the "forthcoming exposure of your malfeasance & corruption" would cause Republicans to abandon him in 2019.

MARCH 25, 2019: After the Mueller Report turned out to contain no damning revelations about Trump and Russia, Brennan conceded to the MSNBC audience, "Well, I don't know if I received bad information, but I think I suspected there was more than there actually was." He added, "I am relieved that it's been determined there was not a criminal conspiracy with the Russian government over our election."

Robert Mueller

KEY BACKGROUND: Robert Mueller was FBI director during a critical time after the 9/11 Islamic extremist terrorist attacks. That's during the same time period in which an inspector general and the FISA court found the FBI had been deceitful in presenting evidence to wiretap US citizens. As a result of the problems, Mueller oversaw the implementation of the new "Woods procedures," which require the FBI to independently verify every single fact presented to the FISA court. If a particular fact is not verifiable, the Woods procedures require the FBI to go back to the drawing board or omit the unverified material from the wiretap application. Mueller, of all people, should have instantly known that the political opposition research "dossier" full of unverified claims, which the FBI presented to justify wiretapping former Trump campaign volunteer Carter Page, violated these strict procedures.

ROLE IN THE NARRATIVE: In May 2017, Mueller—by now retired from the FBI—was appointed as special counsel to investigate Trump-Russia collusion. Attorneys at the Department of Justice who were Clinton supporters and donors were hired to be on his team. There were many leaks from his investigation, including leaks of false information. And when former FBI director James Comey publicly confirmed that the FBI had not verified claims in the anti-Trump dossier before using them as evidence to get wiretaps, it was a *de facto* admission that the FBI had violated its own Woods procedures. However, Mueller and his team stayed strictly on narrative and ignored that. They also declined to investigate the origins of the fake information in the dossier, and they allowed the phones of anti-Trump team members to be wiped after the members were reassigned. They also did not move to prosecute numerous figures who provided the Special Counsel's Office with false information, nor did they prosecute undisclosed lobbying unless it was conducted by Trump supporters.

When Mueller's final report on Trump-Russia collusion didn't deliver what Brennan predicted or Trump's other enemies hoped for, Mueller held a news conference in which he still managed to deliver some negative spin against Trump and took no questions from the press. In July 2019, he testified to Congress about his report. He appeared confused at times and fairly uninformed about the details of his own investigation.

John McCain

KEY BACKGROUND: After his unsuccessful run for president in 2008, Senator John McCain became a chief Republican critic of Trump in 2015, smearing Trump supporters as "crazies." It got intensely personal when Trump hit back, questioning McCain's status as a Vietnam War hero.

ROLE IN THE NARRATIVE: McCain regularly criticized Trump in the press, in Congress, and behind the scenes. He and his associates were among those who met with foreign officials who claimed to have dirt on Trump, and they took steps to deliver the unverified anti-Trump political opposition research "dossier" into the hands of the FBI.

James Comey

KEY BACKGROUND: James Comey manned the FBI during the bungled probe into Hillary Clinton's mishandling of classified information. His FBI also declined to collect the Democratic National Committee (DNC) server in 2016 after the DNC announced that Russians had hacked into it. His agency gave Mrs. Clinton a pass when her representatives erased subpoenaed documents, wiped a relevant computer server, and destroyed Clinton's old mobile devices by breaking them in half or smashing them with a hammer. He also awarded her top aides immunity from prosecution though getting nothing in return.

ROLE IN THE NARRATIVE: Comey worked under President Trump for a short time, collecting notes and material to use against him while assuring Trump three times that he wasn't under investigation. Comey privately briefed Trump about some of the material in the anti-Trump dossier without disclosing that it was political opposition research collected from Russians and funded by the Clinton campaign. Comey's critics theorize that Comey briefed Trump in order to give "news value" to the questionable dossier so that CNN and other news outlets would report the unverified scandal publicly. Until that point, news organizations had declined to publish the unproven and salacious allegations in the dossier. Trump fired Comey in May 2017, after which Comey anonymously leaked negative information about Trump to the *New York Times* through a third party. Comey wrote a book and joined Brennan and Clapper as a frequent critic of Trump. Comey was recommended for possible prosecution after the inspector general found he committed multiple violations during his anti-Trump acts. However, the Justice Department declined to prosecute him. Comey campaigned for Trump's reelection defeat. Some of Comey's tweets include the following:

On July 15, 2019:

This country is so much better than this president. And next year we have a chance to prove it.

On July 18, 2019:

With our voices and our 2020 votes, we must send Donald Trump and his mob back to their dark corner.

On July 27, 2019:

Millions of 2016 Trump voters are not racists. Now those Americans need the strength of character to resolve they will not again vote for someone who clearly is. It's not about judges or taxes. It's about who we are as a nation.

Nobody wants our intel agencies to be used like the Stasi, the secret East German police in the 1950s that spied on and terrorized the country's own citizens for political purposes. The prospect of our own NSA, CIA, and FBI becoming politically weaponized on a large scale was almost unthinkable just a decade ago. Many Americans might have heard about FBI director J. Edgar Hoover's antics and abuses through the early 1970s, but thought such tales were from the past. Now a new generation is coming of age relatively numb to the notion that the US government improperly spies on its citizens—not for the purpose of protecting us but to exploit the information for other purposes. Thanks to the narratives that distract us and keep us looking elsewhere, the unthinkable has become tolerated. Acceptable. Expected.

The tactics of these men, a complicit media, and a segment of the public who wanted to believe the worst about Trump together made the false Trump-Russia story one of the most impactful and effective narratives of our time—and one that has contributed mightily to the public's disintegrating confidence in the media.

The Horowitz Report

On Monday, December 9, 2019, Department of Justice inspector general Michael Horowitz released the long-awaited internal review of the FBI's behavior—or misbehavior—in spying on the Trump campaign in 2016 and 2017.

For anyone who had the time and energy to digest the more than four hundred pages, the findings rank as one of the biggest and most outrageous news stories of our time, cataloging wrongdoing committed by trusted government officials. Any objective news organization that independently evaluated the material would have to agree.

However, well before that release, The Narrative had already been quietly crafted and distributed to the national media. And the media dutifully complied by delivering in a big way.

The news coverage demonstrates typical Washington, DC, tactics in action. Here's how it works: A given interest is worried about what a report or investigation will say, so it launches a well-organized advance campaign to spin expectations and shape how the media report the information and how the public consumes it. To develop the plan, meetings are quietly convened among operatives in public relations, crisis management, and global law firms. Talking points are distributed to "analysts," think tanks, politicians, and political operatives. Obviously, the media play a critical role in executing any successful strategy. They must be willing to accept anonymous "leaks" of information, even when anonymity cannot be journalistically justified. They must be convinced to uncritically buy into propaganda as "the truth" and present it as such. They must book the handpicked, slanted "analysts" to appear on their news programs and quote them in news articles to reinforce the point.

The big takeaway of the Horowitz Report—claimed the media the week before the report was actually released—was that the anti-Trump FBI agents had no political bias. This supposed conclusion was laid out as a narrative and blared across the news before the report became public. We were told the only question that mattered was not whether government agents were guilty of doing anything wrong, but whether they were politically motivated. Classic misdirection.

If the leaks to the media were to be believed, the Horowitz Report was a bust for Trump and those who had suspected our intel officials of improper behavior. And, the media said, it was vindication for former FBI director James Comey, who took the opportunity to publicly crow that he had been exonerated. Driving home The Narrative, the *Washington Post* published an op-ed by Comey in which other left-leaning media, such as The Daily Beast, noted Comey had "spiked the football." You know, like a touchdown in the end zone. *Game over!*

The problem is the spin was far different from the facts. That became apparent when the actual report was released—but it was apparent only to those who took time to pore through it. Many did not. Numerous journalists and analysts figured it was not necessary to read the report itself. After all, we'd all already been told what was going to be in it!

The scope of the disinformation became fully clear when Congress later questioned inspector general Michael Horowitz about his findings. In his testimony, Horowitz acknowledged the report did not exonerate Comey, as Comey had tried to claim. Quite the opposite.

Senator Lindsey Graham, a Republican from South Carolina and head of the Judiciary Committee, asked the operative question: "Former FBI director James Comey said this week that your report vindicates him. Is that a fair assessment of your report?"

"You know, I think the activities we found here don't vindicate anybody who touched this," Horowitz replied.

In fact, the Horowitz Report revealed the FBI had committed a shocking number of egregious missteps and abuses in its Crossfire Hurricane probe examining Trump's supposed connections to Russian president Vladimir Putin. *How could the earlier headline have been that there was "no political bias"?* Horowitz explicitly acknowledged political bias could be among the explanations for the serious misbehavior he uncovered on the part of intel officials.

Because of the outrageous errors and abuses Horowitz discovered, including an FBI lawyer who later admitted he doctored a document to get court approval to wiretap Trump campaign associate Carter Page, a wholesale review of the FBI's wiretap applications was ordered. Ultimately, the Department of Justice and the Foreign Intelligence Surveillance Court had to acknowledge that at least two of the Page wiretaps were improper and invalid due to the FBI's actions, and other wiretaps also showed evidence of FBI lapses and abuses.

Of course, all of this real information was secondary to the slanted narratives that had already plastered the public landscape for a week. The ones that claimed the FBI had no political bias. That the FBI hadn't "spied" on Trump during the campaign. And that the FBI had had every reason to open its ill-fated investigation into Page.

There was an honest burst of outrage from the left-leaning *Rolling Stone* once it was learned what the Horowitz Report really found, compared to The Narrative distributed in advance. The pop culture magazine fumed, "Holy God, what a clown show the Trump-Russia investigation was."

"I hope Carter Page gets a lawyer and sues the hell out of the Justice Department and FBI," said Senator Graham during congressional hearings two days after the report was released.

In short, bad players in the FBI and the Department of Justice targeted the wrong guys and violated their constitutional rights. They used intelligence tools and government authority to go after a political rival and his campaign in a way that there is little historically with which to compare. In a heightened state of political paranoia or political bias, key intelligence officials tossed aside the very rules designed to prevent the abuses they committed. They proved to be so distracted by their zeal to connect the Trump campaign to Russia that they missed the forest for the trees. They were diverted for more than a year by the folly that was Crossfire Hurricane. They spent tax money traveling to foreign lands, wiring agents to approach and record Trump campaign associates, violating the privacy of multiple US citizens without proper justification, and, in at least one proven case, doctoring documents to hide the misguided nature of their pursuit.

In the end, FBI officials claimed they still had every reason to open their investigation into then candidate Trump and others around him, despite their misconduct and the outcome. It felt a bit like a cop getting caught trafficking children for underage sex but saying he had a noble reason: to try to identify pedophiles.

Rarely would I argue the media owe a full-fledged apology to someone for their faulty news coverage. But this case screams for corrections and admissions of mistakes. Apologies should have been offered to President Trump, Carter Page, and the other campaign associates who were improperly targeted.

Instead, many in the media just dug in. The Narrative must be protected at all costs—even in the face of truth and contradictions.

Those who might have watched Horowitz testify before the Senate on December 11, 2019, heard Democrats and Republicans ask questions and could form their own conclusions about what had happened. But The Federalist noted that left-leaning outlets seemed to want to keep their viewers in the dark. "CNN and MSNBC stopped following the IG [inspector general] hearing after about 30 minutes, and both refused to cover the opening statements by Sen. Lindsey Graham, R-S.C. The decision does not align with the recent live hearing coverage standard both networks have held for the last few months, giving endless air time to the impeachment hearings lead [*sic*] by Rep. Adam Schiff, D-Calif, and Rep. Jerry Nadler," wrote the conservative online news site.

Barr Corrects the Record

The prevailing media narrative about the Horowitz Report—that the big takeaway was "no political bias" on the part of the FBI—was so off base that two other federal investigators still looking into FBI misconduct took an unusual step. They went public.

Attorney General William Barr and US Attorney John Durham issued public statements indicating they disagreed with some of Horowitz's conclusions, as well as some of the reporting on the whole matter. Over the next forty-eight hours, Barr gave two nearly identical interviews in which he attempted to supplant The Narrative with his own narrative. The interviews were an unusual information dump for an attorney general who hadn't previously proven to be particularly chatty, especially when it came to his ongoing probe of FBI behavior in the Trump-Russia case.

Here are some of the key points Barr made to NBC News and in an interview with the *Wall Street Journal*'s CEO Council. (All of the information and quotes are from Barr.)

- The FBI did spy on the Trump campaign.
- It was a "travesty," and there were "many abuses."
- US intelligence resources were used to investigate the opposing political party.

- The evidence to start the probe was "flimsy" from the start.
- If the FBI's goal had truly been to protect the election from Russian interference, agents would have given the Trump campaign a defensive briefing.
- The inspector general did not rule out improper motives on the part of FBI officials.
- From "day one," the FBI investigation generated "exculpatory" information and nothing that corroborated Russia collusion. Yet the FBI didn't inform the Foreign Intelligence Surveillance Court, which approved four wiretaps of former Trump campaign volunteer Carter Page.
- The FBI used unverified and allegedly doctored information to get the wiretaps.
- A major Steele source told the FBI in January 2017 that the information he'd provided against Trump was no more than "supposition" and "theory."
- "It was clear the dossier was a sham." Yet the FBI didn't tell that to the court and continued to get wiretaps based on the dossier. Further, the FBI falsely told the court that Steele's source had proven to be reliable and truthful.
- The whole Russia collusion hype was a "bogus narrative hyped by an irresponsible press" that in the end proved to be entirely false.
- "There was a failure of leadership" by FBI director James Comey and former FBI official Andrew McCabe.
- The IG noted that the FBI's explanations "were not satisfactory."

This insight into Barr's thinking on a seminal investigation of our time was arguably even bigger news than the Horowitz Report. But it didn't prompt similar headlines.

As the days stretched on and several reporters did examine the actual inspector general report on the FBI's behavior in wiretapping Page, some important off-narrative journalism was committed. Glenn Greenwald of the website The Intercept rightly took to task the shocking media misreporting and highlighted the astounding implications of what Horowitz said the FBI had done. As he put it:

> *If it does not bother you to learn that the FBI repeatedly and deliberately deceived the FISA court into granting it permission to spy on a U.S. citizen in the middle of a presidential campaign, then it is virtually certain that you are either someone with no principles, someone who cares only about partisan advantage and nothing about basic civil liberties and the rule of law, or both. There is simply no way for anyone of good faith to read this IG Report and reach any conclusion other than that this is yet another instance of the FBI abusing its power in severe ways to subvert and undermine U.S. democracy. If you don't care about that, what do you care about? . . .*
>
> *But the revelations of the IG Report are not merely a massive FBI scandal. They are also a massive media scandal, because they reveal that so much of what the U.S. media has authoritatively claimed about all of these matters for more than two years is completely false.*

But for those not watching closely and carefully, and for people who are not wise to the slanted ways of some in the media, they still have a fuzzy picture of these historic news events, mistakenly believing the FBI was exonerated of wrongdoing, that Trump associates are guilty of something involving Russia, and that any information to the contrary is just political spin.

This is one lesson to keep in mind every time you consume news. How many news events are spun in a similar fashion? How much of what you see is slanted in a way that may be invisible to the casual observer?

CHAPTER 8

CNN: The Cable Narrative Network

"I can't watch CNN," says onetime Cable News Network standard-bearer Lou Waters.

A lot of former CNNers said much the same—unprompted—when I told them I was writing about what has become of our alma mater. Waters, once an anchor at CNN, doesn't mind being quoted by name.

"I can't watch any of 'em," he continues, referring to cable news channels in general. "There's no news anymore on cable television, which is what CNN was invented to provide. It's heartbreaking in a way. And mind-numbing. A threat to everything I grew up with in the news business. I spent a whole career in the news business, and now it's being denigrated. Minimized by false equivalencies between opinion and news."

In short, CNN has become Cable Narrative Network, establishing or carrying water for the political narratives of the moment, almost always politically to the left, unabashedly and without shame. Today, many people consider CNN, along with MSNBC, to be the cable news counterpoints to Fox News and conservative narratives. The difference is that Fox News was well defined as conservative leaning from its inception. The transformation of CNN from a relatively unbiased news source into the notoriously slanted vehicle that it is today has to be the most remarkable devolution in our industry that I can think of. It is also a deeply personal one for me and many longtime colleagues who worked at the old version of CNN.

Many viewers remember Lou Waters for his full white head of hair and classic good looks, resembling the iconic TV news character Ted Baxter. Lou always had the intellect and sensibilities of a thoughtful, seasoned journalist. From 1990 to 1993, Lou and I sat next to each other on the main CNN set every weekday at five o'clock Eastern Time to co-anchor a news program called *Early Prime*. We became good friends. I was twenty-nine, CNN's youngest anchor at the time. Lou was in his fifties and somehow patient enough to put up with a novice like me from local news.

It is hard to imagine in today's politically charged news environment, but during all the time Lou and I spent together—hours anchoring, in meetings, in the makeup room, at dinner with each other's families—and in the three decades since, I have never thought much about what his political leanings might be. It wasn't relevant, it didn't come up, and it didn't matter. As far as I know, he felt the same way about me.

As he speaks to me all these years later on the phone from his home in Oro Valley, Arizona, it occurs to me that I still have no idea where he stands on politics. Except that he thinks it has no place in news.

"Remember at CNN, our goal was 'the news was the star'?" he remarks wistfully. "I don't regret a bit of what we did. I'm just very disappointed in what it turned out to be."

"How would you describe CNN today?" I ask.

"When I do catch occasional glimpses, I call it the all-panel network. If it weren't for the panels, we wouldn't be able to see how we feel about things," he says sarcastically. "Walter Cronkite would roll over in his grave."

Next, Lou talks in a stream-of-consciousness fashion that sounds as though he's been bursting at the seams to say what he thinks about CNN and today's sorry state of the news. "It depresses me, actually. I yell at the TV screen. Back in the day, we did very little politics. The first program was *Crossfire*, a thirty-minute show in early evening that was point-counterpoint. That was our politics. The rest was regular news—what was going on in the country and the world outside of Washington, DC."

I get an equally dismal assessment from a former top CNN executive who describes himself as socially progressive and does not wish to be quoted by name. "Nobody who watches CNN thinks they're anything but liberal, and I think their content shows it," he says. "Too many of their shows spend their whole time attacking conservative agendas. It's too easy to attack Trump, but they don't put the same energy into the progressive side."

Former CNN world affairs correspondent Ralph Begleiter adds, "In routine viewing, CNN does not give you a comprehensive picture. It's very narrowly focused on the political battle in Washington, DC. I'm not saying that's not important, but it's not a picture of the world today."

"You almost never see a story. They do panels," says a top TV network news executive, echoing Lou's observation. The executive calls Trump a "disaster in many ways" but criticizes what he sees as CNN's biased approach to covering the president. "It's obviously a very hard place to be, the White House, the presidency, your skin has to be so thick and [Trump] suffers from having the thinnest skin of almost anybody. But it's outrageous so much of the reporting really comes off as anti-Trump. . . . I find it shocking so many people on the air are not at all concerned about showing their disdain for the president, and they scoff a lot. I find that to be hard to take." The executive goes on to say, "I expect it at MSNBC, and they don't pretend at Fox. They're pandering. But CNN?"

Nothing better demonstrates the vast difference between CNN back in the day and CNN now than how anchors wrap up live news events such as a presidential speech. The "wrap-up" is the part of the news where the anchors are seen on the set at the end of the speech. As a CNN anchor, I used to provide a simple, factual summary and a bit of context. For example, I might say something along the lines of "That was the president speaking for the first time since the hurricane in Texas. He announced plans to tour some of the damaged areas tomorrow. When asked how much money will be committed to residents for hurricane aid, he said federal officials will be consulting with the Texas governor in the coming days for an assessment."

Contrast that with the wrap-up CNN's Anderson Cooper gave in July 2018 after President Trump's news conference following a summit with Russian president Vladimir Putin. "You have been watching perhaps one of the most disgraceful performances by an American president at a summit in front of a Russian leader, certainly that I've ever seen," said Cooper. No pretense of being factual or neutral.

This is the sort of presentation that CNN president Jeff Zucker not only tolerates but encourages. Some news industry insiders say it's a bottom line–driven strategy. "Zucker is money hungry," says a news executive who once worked with him.

In conducting research for this book, I pursued an interview with Zucker to get his take and hear his vision for CNN's future. I also asked CNN's press office to connect me with representatives who could answer questions and give a positive assessment of the news organization. However, Zucker would not agree to an interview. The press office would neither comment nor refer me to anyone else to talk to.

CNN was the first twenty-four-hour news network. Its dramatic transformation from "just the facts" to Narrative Central largely tracks with the death of the news as we once knew it. When I jumped on board in 1990, it was the first and only national channel that was all news, all the time. I feel lucky to have had the experience of working there during its golden years. The start of that adventure came when I was working for the local CBS-TV affiliate in Tampa, Florida, WTVT, and CNN vice president Paul Amos called me up for a job interview at CNN headquarters in Atlanta.

CNN, Circa 1990

As I'm escorted around CNN world headquarters, my head is spinning. The multi-floor offices and studios are modern and spacious. Like cogs working in a well-oiled machine, the giant staff of newshounds working in an open-plan newsroom churn out a nonstop flow of information. Off to one side is a glass wall between the newsroom and the set where the anchors read the news. It's epic to see, in person, the operation I'd long watched on TV from afar.

CNN executive Bob Furnad rushes me upstairs into a room equipped with mirrors lined with Hollywood-style light bulbs. He gives instructions for the makeup artist to make me TV ready. She swipes on an extra layer of foundation, blush, lipstick, and mascara and touches up my hair—big hair with permed curls (*en vogue* at the time, I promise). After a few minutes, Furnad returns to the makeup room and hurriedly leads me into a small studio without saying more than a few words. He tells me I'm going to do a "beeper." I realize it's part of my audition. I've never heard the term "beeper" before, but I don't want to ask what it is. Furnad barks out a few sketchy facts, something along the lines of "There's been a plane crash in New York at LaGuardia Airport. There's a witness on the phone. That's all you know." He then tells me, "*Go!*" I quickly deduce that a "beeper" is CNN's term for an anchor's live, audio-only interview with a newsmaker or someone on the scene of a news event—sort of a telephone interview but with the anchor on camera. (We called them "phoners" in local news.) In other words, Furnad wants to see how I'll handle a telephone-type interview in a mock breaking news situation. I look into the camera, trying my hardest to feel as though I'm really on live television while ignoring the fact that I'm being critically examined by national news network managers. I begin by setting the scene for the "viewers"; then I "interview" the "eyewitness" (played by Furnad, who is calling out responses as if he's at the scene on the phone). Afterward, Furnad provides me absolutely no discernible feedback. No smiles or friendly pats on the back. He hustles me off to Amos's office, and Amos walks me down the hall to the office of CNN president Burt Reinhardt.

Reinhardt looks small in his chair. He is seventy—seems old to me at the time. He appears to have a slightly friendly twinkle in his eye as he measures me up. I don't know it at the time, but he's a bona fide legend. A World War II combat photographer whom Ted Turner tapped to help launch CNN in 1980. Reinhardt is about to retire. Anyway, I have apparently passed the day's test, because I get offered the job and luck into a seat as a primary CNN daytime anchor.

One of my first assignments is filling in as co-anchor on the flagship six o'clock newscast with lead anchor Bernie Shaw. I'm eventually assigned to anchor more CNN newscasts than any of my colleagues: the programs at noon, 2:00 p.m., 4:00 p.m., and 5:00 p.m.

Not long after my start date in August 1990, Iraq invades Kuwait, and the first Gulf War begins. CNN's coverage of the Gulf War puts us onto the map in a way that eclipses all previous news events. Our reach and coverage are so far superior to those of any other television news organization at the time that something unprecedented happens. When we have exclusive breaking news of missiles being fired at US forces in the Mideast, broadcast network affiliates around the country dump out of their normal programming and begin airing *our* feed on their channels. Incredibly, that means that during the Gulf War, viewers anywhere turning on CBS, ABC, and NBC would see *our* live coverage on CNN!

Back then, I think we were just about as "just the facts" as it was possible to be. We news anchors wouldn't have dreamed of slamming political figures or giving editorial monologues about them during our news reports. Most of the news we aired wasn't Washington, DC–centric or even political in nature.

During my tenure, I recall CNN had one main political news program. It was a half hour called *Inside Politics*. And the only political debate–type program was the nightly half hour called *Crossfire*, which Lou Waters mentioned earlier. The hosts at the time, conservative Robert Novak and liberal Michael Kinsley, were famous for extensively preparing their evidence and arguments in a way that seems completely foreign to TV talking heads now. Equal consideration was given to conservative and liberal viewpoints regarding the news of the day.

"The news," at least back then, encompassed much more than one or two political narratives. Besides our regular news programs, we had a daily Hollywood-centered entertainment news program called *Showbiz Today*. There was a daily lunchtime talk show called *Sonya Live*, which explored social and psychological issues. It was billed as "intelligent talk for intelligent people." For a time, I anchored *CNN International Hour* at midday, where we reported on news from around the globe. On that program, I might interview Pakistani prime minister Benazir Bhutto one day, Libyan leader Muammar Gaddafi the next. I co-anchored *Early Prime* with Lou at 5:00 p.m. Eastern Time. That program focused heavily on domestic news of interest to America's aging baby boomers. This meant stories on a wide variety of topics, including finances, Congress, retirement, education—you name it. We had a nightly business news program, *Lou Dobbs Tonight*. And there was *Larry King Live*, which featured interviews five nights a week with all kinds of people, including celebrities, athletes, and politicians.

My point is: the universe of news we covered seemed much bigger then. The breadth of topics was a more diverse, informative mix. We weren't responding to or promoting nonstop political narratives.

I think part of the magic of it all was that nobody had to tell us—at least nobody told *me*—that CNN defined itself as a fact-based information operation. We just knew it. When I was hired, there were no briefings or orientations about keeping my personal opinions out of the news. I just understood that's the way it was.

I give a great deal of credit for all of that to CNN's founder, the ultra-liberal billionaire Ted Turner. Nicknamed "The Mouth of the South," Turner was opinionated. But he understood that the mission of his news network would be undermined if the news product were not perceived to be generally neutral. As tempting as it might have been for him to turn his invention into a twenty-four-hour-a-day personal editorial messenger for his chosen liberal causes, he didn't do that. Turner was pushed out of CNN in 2001 after Warner and America Online acquired it. When interviewed in 2018, Turner politely expressed distaste for what his creation has become.

"I think they're stickin' with politics a little too much," Turner told former ABC anchor Ted Koppel in the interview. "They'd do better to have a more balanced agenda. But that's, you know, just one person's opinion."

I think Turner's remarks reflect the understated tact of a southern gentleman. Other people affiliated with CNN during its heyday offer more direct criticism of today's version. Some of them think CNN has sold its news soul to chase ratings and money.

"Ted always said, 'Let's make the news the star,'" David Bernknopf recalls. Bernknopf was a founding employee of CNN. He was a colleague of mine and became a friend during my years there. He stayed on at the cable news operation for years after I moved on to CBS News. He eventually served as CNN's first vice president and director of news planning. In one of life's surprising and pleasant turns, we reconnected professionally in recent years when I was able to convince him to join me to become an investigative producer on my Sunday television news program, *Full Measure*. That was a quarter century after we first met as young journalists at a young news network, then celebrating its first decade.

"Clearly in 'the good old days,' as a lot of original CNN people look at it, we didn't have highly paid stars. We couldn't afford them," Bernknopf says. "We didn't have an internal history. In our rush to get on the air, maybe we couldn't think as much about some of these things. We were just trying to get the news on the air. When you have a bunch of people who aren't stars, who don't have the power as anchors or reporters to craft an image, you just go out and do your job."

At the start of CNN, Lou Waters tells me, the anchors used to be considered so unimportant that the network didn't even want them to identify themselves to viewers. "We weren't even allowed to say our names in the beginning," he says. When Lou's wife gave birth to twins, there were strict instructions that viewers were not to know that personal detail. "No one is to mention anything about this on TV," Waters says CNN executives told the other anchors at the time. "That's how un-personality-driven CNN was."

What a contrast to today!

Begleiter says the personality-driven culture at CNN is one of the biggest differences he sees between today's CNN and the vintage version. He divides the way he talks about CNN into two periods: the first twenty years (1980 to 2000), which he describes as the "pre-digital, pre-political age," and the twenty years since (2000 to 2020)—the "post-digital, post-political age."

"The 'pre' is exemplified by the fact that the CNN anchors were not all that well known and that nobody made any assumptions about what their political points of views were," Begleiter observes. "And in the 'post' period, it's fair to say that almost the first thing people think of when they think of the on-air personality is what their political stance is, what's their political view. I think that's a watershed change, not only at CNN."

Another former prominent CNNer told me he has a hard time accepting the news network's current approach. "There's a lot of showboating going on on television at every level, now," he says. "News *was* the star. And now the star is the star. 'Hey, look at me!'"

Jim Acosta

Numerous CNNers I spoke with brought up the name of CNN White House correspondent Jim Acosta. I worked with Acosta at CBS News, where he was employed as a reporter from 2003 to 2007 prior to his time at CNN. We were based in different cities, so we didn't really know one another personally. From what I knew, he was well regarded, if not particularly well known, and considered a decent guy to work with. But at CNN, his openly biased anti-Trump tilt has come to symbolize how much has changed about CNN and the news. So how and why did he make a name for himself at CNN in this way—among both fans and critics?

Probably the biggest incident that made Acosta the focus of the news rather than a reporter of it happened on November 7, 2018.

President Trump called on him at a White House news conference and answered several of his confrontational questions. Then, when Trump tried to move on to the next reporter, Acosta tightly clung on to the White House microphone, refusing to give it up and brushing away a White House press assistant who tried, unsuccessfully, to retrieve it.

After the high-profile incident, the White House temporarily suspended Acosta's "hard pass." The decision was widely attacked by media groups, and CNN filed a court case to object. CNN claimed the revocation of the pass violated Acosta's First and Fifth Amendment rights.

By way of background, a hard pass is what grants journalists "anytime access" to the White House press space. No journalist is automatically entitled to such a pass. Each applicant's name is submitted by his news organization, and he must meet eligibility requirements. To get the pass, an applicant's primary job must be covering the White House as a journalist, and he must clear a background check. News organizations are not permitted an unlimited number of hard passes. Reporters who do not have a hard pass can apply for admittance to the White House to cover an event any day with case-by-case approval.

The judge hearing CNN's request for a temporary injunction to get Acosta's pass returned, a Trump appointee, said the White House should let Acosta back into the briefings while the case worked its way through court. The White House did so, noting that it would be developing new rules designed to maintain order and fairness at the press briefings.

It was about that time that the White House halted the traditional press briefings altogether. On January 29, 2019, President Trump tweeted:

The reason [White House press secretary] Sarah Sanders does not go to the "podium" much anymore is that the press covers her so rudely & inaccurately, in particular certain members of the press. I told her not

to bother, the word gets out anyway! Most will never cover us fairly &
hence, the term, Fake News!

One could say that Jim Acosta inadvertently caused the White
House to cancel a fifty-year press tradition.

It robbed some reporters of the opportunity to grandstand and
get clips of themselves published on the news and passed around
on social media. But it didn't seem to make much difference in the
information flow. Instead of the canned briefings, White House
reporters got frequent, direct access to the president himself, who
stopped more often to talk to the press and took longer Q-and-A
sessions than anyone before him. During the coronavirus crisis,
Trump personally faced reporters almost every day, often for two
hours at a time, taking their questions.

Meantime, it was soon revealed that at the same time Acosta was
grabbing the mic at the White House and becoming the story, he was
also writing a book criticizing Trump titled *The Enemy of the People:
A Dangerous Time to Tell the Truth in America*. That's something that
would never have been permitted at the old CNN.

Although the news media and journalist groups almost exclu-
sively supported Acosta publicly, I was surprised that so many journalists I spoke to—who do not like Trump and consider themselves
liberal—said they felt Acosta was out of line.

The former CNN executive who describes himself as progressive
brought up the Acosta book deal when I spoke with him about our
former place of employment.

"When I heard Jim Acosta had written a book about Trump while
covering the White House, I thought, 'This is nuts!'" says the official.
"The idea of a White House correspondent writing a book about their
experience at the White House in the middle of that administration
while they are still covering the administration seems nutty to me.
How do you cover an institution when you criticize how an institution
is treating you? What becomes more important in that situation is
generating heat, not light. It's more important that Jim Acosta is not

getting along with Sarah Sanders or Donald Trump or his communication shop than it is to get useful information while on the job, because that fits into a CNN narrative that 'We're the ones who are tough on Trump.'"

A former top CNN official, who describes himself as liberal and who politically opposes President Trump, also brought up Acosta and his antics. "If I were chairman of CNN, I would call the White House correspondent and say, 'Nobody elected you, and you're not there to fight with the president of the United States. You're not there to battle with him.' They make [Acosta] a folk hero, and I don't think he is, and I do not think others think so, either."

But another top network news executive expressed a different view. He told me Acosta is doing the right thing. "Acosta does a good job," he says. "When you make it about you, yes, I have a problem. But the White House job has always been about holding the president accountable whenever you have a chance to. Particularly with this president because he's such a lightning rod and has such thin skin. That's the job; the one chance you get [as a reporter] is when he's standing up there and you've got the microphone." The executive adds, "[Trump] should have more fun with people like Jim Acosta."

Whatever your view, there is no disagreement that at the old CNN, a Jim Acosta would not have been possible. A reporter publicly expressing animosity toward, or really any strong opinions about those he covers would have been admonished if not summarily dismissed from his job. It is the death of the news as we once knew it that has made this new dynamic possible.

"Zucker could do a better job by far of reining in his anchors," says a former top network news executive who describes himself as "liberal leaning."

A former CNN executive remarks, "We've decided that commentating is more important than news. I left cable because I couldn't understand the screaming. I fought the fight to do quiet, straight journalism. And it didn't win at CNN."

"Maybe some people did call us boring," says Bernknopf, speaking of the old days. "That was one of the criticisms, and maybe that was fair. But for my mind, I'd rather be boring than a lot of what I see now. CNN management has decided its niche is going to be, while Trump is president, it is going to cover every single small, medium, or large development like it's a nuclear bomb."

Putting the Blame on Fox

To the extent that so many people in the news business seem to think CNN has lost its way, it took me aback that so many of them blame Fox News. In their view, Fox News pioneered the pandering, biased cable news model, and the conservative channel's success paved the way for CNN and others to seek to become its liberal equivalents.

"Being overtly biased is a concerted effort [on CNN's part]," says a news official who served as an executive at several broadcast networks and CNN prior to Zucker's tenure. He knows Zucker. He also knew Fox founder Roger Ailes before Ailes started Fox News in 1996.

"Back in the day, Roger called CNN the 'Clinton News Network.' CNN had a liberal reputation but was actually fair," says the executive, at the time a high-ranking official at a broadcast news network. "Ailes said to me, 'Seventy percent of the country, maybe more, think they want to watch unbiased news and not get all muddied up in political battles. But thirty percent are underserved. I'm going to reach out and be their channel. I'll have a third of the news viewing audience dedicated to me, and the rest of you can split the rest. You'll have slivers, but I'll win.'"

That executive goes on to say that when Fox News proved to be an unexpected and remarkable success, it gave Zucker big ideas. At the time, Zucker headed up MSNBC. "Zucker came to me and said, 'We want MSNBC to be the Democrats' alternative to Fox.' I said, 'That's not what I want to do.' Zucker said, 'But that suits your politics, doesn't it?' And I told Zucker, 'If I'm working, I don't care about my politics. I want to do great programs. A mark of that is to be un-biased.'"

Zucker became president of CNN in 2013.

"Jeff is what you would call a day trader of news," says another news executive who worked with Zucker when they were both at NBC News. "He senses what the story of the moment is and wants to jump on it with everything he's got. Trump gave him the opportunity for better ratings at CNN, a first taste. Then it became a nightmare. As long as there's Trump, they have high ratings. Well, they're actually shitty ratings, but there is enormous ad revenue coming in, and they have strong pockets of support. They're losing in ratings to MSNBC but making more money than they can count."

"Zucker is making a business decision more than a personal ideo-logical decision," another of Zucker's colleagues tells me. "When they look at the numbers, they're driven by panels about Trump misbe-having, not by news reports from around the world. Zucker gave di-rectives when the [Trump] impeachment started that 'We should be all over this in a big way,' and that was misinterpreted as 'We should bring Trump down.'"

The executive adds that he thinks CNN is "missing a huge oppor-tunity" to draw viewers in with news coverage utilizing its amazing global infrastructure. "Pandering to anti-Trump sentiment only hurts them with anyone who wants them to be even."

Again, Zucker and CNN declined my repeated requests for inter-views and information.

Former CNNer Begleiter blames Ailes for pioneering the seem-ingly endless parade of analysts, panels, and political narratives that is much despised by many journalists. "Roger Ailes knew it was cheaper to talk about what other people are reporting rather than do the reporting," he says.

As negative as some of the reviews of today's CNN seem to be, the analyses aren't all bad.

Despite his critical observations, Begleiter also has some kind words for CNN. He says that even today, nobody does breaking news better. He adds that CNN International, which is seen in foreign countries, remains "truer to its original self of factual global news reporting." And he doesn't view CNN as being of the same politically biased ilk as Fox or MSNBC, which, he says, "feed their audiences." CNN "is still better than anything else," he tells me.

One former CNN executive told me that although he disagrees with a lot of CNN's current strategy, he still watches. "I watch because they have certain features I really enjoy," he says. "On election night, I think John King is great. I'm a friend and fan to Jake Tapper, although I kind of wish someone would tell him to rein it in at times. And when I really want to be entertained, I watch Don Lemon because he's so stupid. Well, he's not stupid, he's a journalist, but he's as biased as [Fox News commentator Sean] Hannity is."

More than one news executive told me that CNN will have to change itself again when it no longer has Donald Trump to kick around and when Trump is no longer making it famous and infamous with his "Enemy of the People" rhetoric.

When that time comes, will CNN still focus on presenting narratives and slanted information rather than sticking to the facts? Will that bring in enough accolades and support for it to continue onward? Or will viewers decide they're weary of the rhetoric? Is there a chance CNN's leaders could see the value in returning to a more neutral and factual tone?

"My question is, I just wonder what CNN is going to do when they don't have Donald Trump anymore," says a former top CNN executive, "because they've chased away their loyal audience. They've ruined their reputation."

Another former CNN executive agrees. "Post-Trump—how does CNN tell their audience, 'Now we're a news network again'? Jeff will retire, and it will be somebody else's nightmare," he predicts.

In March 2019, Zucker added "Chairman, WarnerMedia News and Sports" to his "CNN News President" title.

Pundits and Polls: Hard to Believe

Poor Bernie Sanders. He was on a roll prior to March 3, 2020, Super Tuesday, the day fourteen states hold their presidential primaries. But on the eve of the big vote, lesser Democrats dropped out of the race and endorsed Sanders's rival, Joe Biden. Biden, in turn, made what was hailed as one of the most surprising comebacks in modern politics, going from near zero to hero and taking the lead in the primary delegate count.

On March 8, 2020, Sanders says what I've been thinking. In an interview with George Stephanopoulos on *ABC This Week*, he laments "the power of the establishment, to force Amy Klobuchar, who worked so hard, Pete Buttigieg, who, you know, really worked extremely hard as well, out of the race. What was very clear from the media narrative, and what the establishment wanted, was to make sure that people coalesced around Biden and tried to defeat me. So that's not surprising."

Sanders was a roadside casualty of the fact that pundits, the media, and the news reports about the polls have once again proved wildly off on their analyses. To be fair, he wasn't the only one victimized by slanted news coverage and narratives. Just as the prevailing wisdom never saw a Trump presidency coming, the same players incorrectly declared Biden to be dead in the water early on. So Biden had survived his own bout with The Narrative. Prior to his Super Tuesday surge, fellow Democrat Van Jones of CNN wrote Biden off as "dead man walking." Other Democrats urged him to give it up. The

Narrative was that he could not possibly win the nomination. *Just look at the polls!* But The Narrative and the polls were wrong.

In the end, Biden "rose from the dead" only because it was the media that put him six feet under in the first place. His ultimate success had to be billed as a surprising, remarkable turnaround, or else the pundits and analysts would have to admit that once again, they'd got it all wrong. And that—they seem unwilling or unable to do. There's a reason for all of this. Just as The Narrative calls upon the news to codify certain story lines, political polls are now widely used for the same purpose. Polls have morphed from providing a snapshot of public opinion at a moment in time into being an indispensable tool used to shape voter opinion.

"Of course people use polls to shape public opinion," polling expert Scott Rasmussen of ScottRasmussen.com tells me. "If you went back twenty or thirty years, there weren't as many public polls, so this polling that is out there now is brought in to shape a number of things."

A lot of people probably do not put much thought into how polls work and their relationship with news organizations. When news outlets or companies commission polls, they get to decide what questions are asked, how they are phrased, and what headlines are chosen from the results. Rasmussen says, "Obviously, the organization paying for the poll can use it however they want. They can select the questions. They can interpret it as they want."

In this way, polls have become essential elements in advancing political horse race narratives.

The bugle sounds! The gates open!

Fifteen full months before the 2020 election, competing poll-related narratives are already in play.

A poll by Quinnipiac University in August 2019 finds that every top major Democrat would beat Trump by at least nine points. Some analysts press the Democrats who are hovering near the 1 percent mark to hang it up. Yet the election is still a lifetime away, in political terms. All those involved seem to have forgotten that Donald Trump was hovering near the bottom at the exact same time in the 2016 campaign. One could extrapolate that early polls cannot be taken as hard indicators of what will come, especially when you consider that polls are often used to further the narrative that things aren't going the way you think they are. *It's anybody's ball game*—(especially if the guy the media are pulling for is actually behind).

For example, in May 2015, a Quinnipiac University poll found that Donald Trump topped the "no way" list among Republicans, with 21 percent of GOP voters saying they would "definitely not" support him. In June 2015, an NBC/*Wall Street Journal* poll put Donald Trump at 1 percent, behind ten Republican candidates: Jeb Bush, Scott Walker, Marco Rubio, Ben Carson, Mike Huckabee, Rand Paul, Rick Perry, Ted Cruz, Chris Christie, and Carly Fiorina. Politico's Daniel Strauss tried to tamp down fears that Trump could actually be a winner. "Whispers of a Trump surge are making the rounds," he wrote. "It might be wise to take a deep breath . . . nationally Trump's polling has been on the decline."

Former New Hampshire Republican Party chairman Fergus Cullen declared there was "no visible grassroots movement for Trump" in New Hampshire. And Patrick Murray, director of the Monmouth University Polling Institute, said, "At the end of the day, it's quite possible that Donald Trump will get 11 percent in New Hampshire, but that might be his cap." (Trump ended up winning the Republican primary in New Hampshire with just over 35 percent of the vote, more than triple Monmouth University's prediction.)

In July 2015, a *USA Today*/Suffolk University poll found Trump trailing Clinton by 17 points, 51 to 34 percent. In September, an NBC/*Wall Street Journal* poll found "The only Republican whom Clinton led by a significant margin was businessman Donald Trump." She supposedly had a 10-point advantage.

Moving closer to the primaries, in November 2015, Nate Silver of the polling site FiveThirtyEight concluded Trump's odds of winning the presidency were "higher than 0 but (considerably) less than 20 percent." In December, an NBC/*Wall Street Journal* poll determined that Hillary Clinton would beat Trump by 10 points. A Quinnipiac

University poll found Clinton would thump Trump 47 to 40 percent. An NBC/*Wall Street Journal* survey determined Clinton would "smash" Trump 50 to 40 percent. Fox News said Clinton would best Trump 49 to 38 percent. Deroy Murdock of *National Review* predicted a Trump nomination would "engineer a Hillary Clinton landslide." He advised that Marco Rubio would be "a far more elusive target for Clinton's slings and arrows."

In January 2016, David Wasserman wrote on FiveThirtyEight that a Donald Trump nomination would "make Clinton's election very likely and raise the odds of a Democratic Senate." He said, "In other words, if you're a member of the Republican Party who wants to win in November, it's basically Rubio or bust." And in March 2016, a just-the-facts analysis of hard data by The Conversation determined Trump would not win enough electoral votes to beat Clinton, ending up with just 236—34 fewer than the 270 needed. (It was off by 68 electoral votes. Trump received 304, 34 more than needed. Clinton actually ended up with fewer electoral votes than the deficit predicted for Trump.)

All of this has contributed greatly to the media's declining credibility among the voting public. Yet there was no mountain of mainstream analysis or criticism of these erroneous polls and predictions. Who *did* get attacked after the surprise results of the 2016 presidential race? Why, the polling group that proved to be the most accurate among them.

Narratives about Polls

A subpart of the concept of polls as narratives is narratives about polls. Especially when a particular poll is off narrative.

If you understand the propaganda efforts that use polls to advance The Narrative rather than as legitimate measures of public opinion, you understand why polls with off-narrative results, even if accurate, must be controversialized. If they were not, people might believe them.

Rasmussen Reports learned this the hard way after it beat nearly every other major polling group in terms of accuracy in the 2016 presidential election. Although almost nobody had Trump beating Clinton, Rasmussen Reports was the only pollster to accurately predict Clinton would win the popular vote over Trump by about 2 percentage points. (As a side note, Scott Rasmussen and Rasmussen Reports severed their relationship years ago and are now separate entities.)

We start in December 2018, just after the midterm elections. A Rasmussen Reports poll shows approval ratings that are more favorable for President Trump than those in other polls. Trump tweets the numbers. That's all it takes for the media pushback to rear its ugly head. The attackers are not only going after Trump; they are also attacking Rasmussen Reports. It is yet another demonstration of how the media have transformed the way they see themselves from reporters of facts to shapers of opinion. CNN titles its takedown of Rasmussen: "Trump's favorite pollster was the least accurate in the midterms."

"Just this week, the President tweeted out a result from his favorite pollster, Rasmussen Reports, that showed his approval rating stood at 50 percent. Rasmussen's polling does not meet CNN standards for a number of reasons including that it doesn't call cell-phones," writes CNN's Harry Enten. Of course, Enten does not mention that CNN's own poll in July 2015 found "[Hillary] Clinton's clearest advantage [among Republican contenders was] over Donald Trump." In fact, at this point, CNN put Clinton's advantage over Trump at 25 full percentage points: 59 to 34 percent. Never mind that. It is not about accuracy, after all. It is about controlling the story line and telling people only that which will convince them to think a particular way.

"The fact that Rasmussen has a better approval rating for the President than other pollsters isn't new," chides CNN's Enten in 2018. "This is why we've seen Trump mention Rasmussen many times." He concludes that "The midterm [2018] elections prove that at least for now Rasmussen is dead wrong and traditional pollsters are correct."

"Dead wrong"? "Traditional pollsters"? CNN and other media seem to be working overtime to further the notion that Rasmussen Reports polls are somehow not to be trusted. To imply that the polling group is using strange, untraditional methods.

In fact, Rasmussen Reports has been polling since 2003 and came out of the gate praised for its accuracy. Slate and the *Wall Street Journal* were among those who noted that Rasmussen Reports was one of the most accurate pollsters in both the 2004 presidential election and the 2006 midterm elections. In 2008, even the liberal outlet Talking Points Memo wrote, "Rasmussen's final polls had Obama ahead 52%–46%, which was nearly identical to Obama's final margin of 53%–46%," making Rasmussen Reports "one of the most accurate pollsters."

In 2012, Rasmussen Reports did suffer a setback. It put Republican Mitt Romney ahead of Barack Obama in the presidential race, "overestimating Mr. Romney's performance by about four percentage points, on average." As a result, a Fordham University ranking conducted by a former Clinton Senate staffer, Costas Panagopoulos, ranked Rasmussen Reports twenty-fourth out of twenty-eight polls in terms of accuracy for that presidential election.

But the 2016 presidential election is a different story. Rasmussen Reports once again comes out on top in terms of accuracy, as I've described. (Interestingly, Fordham's Panagopoulos omitted Rasmussen Reports from his 2016 ranking, when Rasmussen Reports would presumably have been listed at the top.)

That brings us to December 2018 and Rasmussen's 50 percent approval rating for Trump. As an outlier, the poll triggers the familiar media outrage. Philip Bump of the *Washington Post*, Nate Silver of ABC News, and CNN's Harry Enten label Rasmussen Reports the "most inaccurate pollster" in the 2018 congressional midterm elections, claiming it was off by almost ten points. According to the news outlets, Rasmussen Reports "projected the Republicans to come ahead [*sic*] nationally by one point, while at the time Democrats were actually winning the national House vote by 8.6 points—an error of nearly 10 points."

Rasmussen Reports quickly punches back. It accuses the press of deliberately misconstruing its work. It never polled about the House of Representatives, the race naysayers claimed it had gotten so wrong. Therefore, it could not have made a 10-point error in its House projection; it had not made a projection. Instead, as it had done for a decade, the Rasmussen Reports poll asked one question collectively covering both chambers of Congress, the House and the Senate: "If the elections for Congress were held today, would you vote for the Republican candidate or for the Democratic candidate?" That's the question that netted the 1-point advantage for Republicans. Rasmussen Reports argues that the 1-point cumulative edge given to Republicans in the combined House and Senate was hardly a 10-point miss. After all, Republicans lost the House but held on to the Senate, where they actually picked up two seats.

To make the media slant against Rasmussen Reports more obvious, its critics misrepresented the wording Rasmussen Reports used in its polling. As Rasmussen Reports explains, "So eager were [critics] to report about Rasmussen Reports 2018 generic ballot 'failings' that [they] unilaterally changed Rasmussen Reports wording scope in their articles from 'Congress' to 'the House,' thus reaching down in an apparently coordinated fashion for a new historic low in national poll 'analysis.'"

No sooner did Rasmussen Reports defend itself than it came under further attack by Wikipedia editors, who inserted criticism into the Rasmussen Reports Wikipedia page. "Rather than rethinking their methodology," wrote the Wikipedia propagandists, "Rasmussen pushed back against critics after their widely derided miss . . . choosing instead to attack those who discuss the matter and citing 'bad actors' who 'create chaos.'"

Naturally, Rasmussen Reports' counterpoint—its explanation—was not allowed to appear on the Wikipedia page. The Wikipedia editors controlling the page deleted it. *There. The Narrative stands in the rec-ord. Wikipedia says it, so it must be true.*

Though Rasmussen Reports came under rapid-fire attack, you typically will not see such criticism of other polls that are actually wildly wrong. When their results support The Narrative, even when they are incorrect, they are widely defended.

One little-told example of bad results that were not subject to widespread critique came one month before the 2016 presidential election. It's an amazing story of a mysterious polling turnabout.

The story starts in early October 2016. A videotape recorded in 2005 by *Access Hollywood* is leaked to the *Washington Post*. In it, Trump is heard making vulgar remarks and bragging to *Access Hollywood*'s Billy Bush about what women will let men do "when you're a star."

This time, insists the press, *Trump* might *finally be done for!* "An unlikely Bush finally did some damage to Donald Trump: Billy Bush," reads a *Post* headline.

By late October 2016, an ABC/*Washington Post* poll shows Hillary Clinton leading Trump by 12 points. Associated Press releases its own poll the same month showing Clinton ahead by a whopping 14 points!

Then something strange happens.

About a week later, the same ABC/*Washington Post* poll suddenly . . . flips! It shows Trump ahead by 1 point. That's a 13-point switch in one week—in Trump's favor!*

Such a large shift in polling is widely considered implausible. The true popularity of a candidate doesn't swing so drastically in one week without a discernible intervening event. Did all the voters who supposedly were so angry about Trump's words on the *Access Hollywood* tape suddenly decide to forgive and forget?

There's reason to question whether Trump's popularity had really

---

* The swing was prior to FBI director James Comey notifying Congress on October 28, 2016, that the FBI was looking into additional emails pertinent to the investigation into Hillary Clinton's handling of classified information when she was secretary of state.

ever taken the hit that the polling claimed to find. So why did it look that way?

One possible explanation comes three years later in a surprising tweet from President Trump. The tweet, in September 2019, reveals that the Trump campaign had apparently threatened to sue ABC and the *Washington Post* after the late–October 2016 poll showing Trump down 12 points. The campaign viewed the results as malicious and wildly inaccurate. Trump tweets:

> ABC/Washington Post Poll was the worst and most inaccurate poll of any taken prior to the 2016 Election. When my lawyers protested, they took a 12 point down and brought it to almost even by Election Day. It was a Fake Poll by two very bad and dangerous media outlets. Sad!

Was Trump correct to think the polling was somehow rigged? Is that why the point spread suddenly seemed to adjust back to its normal place a week later?

One might ask why Trump waited until three years later to break the news that he'd threatened to sue over the 2016 ABC/*Washington Post* poll. The answer appears to be that he was trying to overtake The Narrative after a September 2019 ABC/*Washington Post* survey showing his approval rating down.

As if on cue, here comes the left-leaning website Vox to defend the ABC/*Washington Post* polling that was negative for Trump: "Trump, of course, has a long history of attacking polls that don't reflect well on him. But his specific claim about 'inaccurate' 2016 polling is simply false." The blog goes on to defend the unlikely 13-point swing in the 2016 ABC/*Washington Post* poll by saying that public opinion had probably changed that much in a week.

SUBSTITUTION GAME: Rasmussen Reports' accuracy in the 2016 presidential race proved better than most but was mercilessly critiqued by the media. On the other hand, Vox defended the implausible 13-point swing in the ABC/*Washington Post* poll because it supported a popular anti-Trump narrative. In fact, Vox itself has a checkered record when it comes to election predictions. The website relies on Sean Illing for political analysis. Illing declared in December 2015 on the liberal Salon website that if Trump faced off against Clinton, it "would not only hand the presidency to the Democrats, it would also [according to a recent Politico survey] 'lead to a Democratic landslide up and down the ballot.'" In truth, Republicans retained their majorities in the House of Representatives, Senate, and governorships. And then there is Vox's Ezra Klein, who authored such misses as the June 7, 2016, blog titled "It's time to admit Hillary Clinton is an extraordinarily talented politician" and the October 19, 2016, foolishness "And it's not just the presidential race. Betting markets now predict Democrats will win the Senate." (Democrats did not win the Senate; Republicans picked up seats.) And finally, there was this prediction from Vox's Matthew Yglesias the day before the election: "The point is simply this: Clinton is clearly up in the polls and is more likely than not to win."

In my view, political election polls are meant to be passive measurements of how a group of voters feels at a given point in time. They are not supposed to be tools to shape public opinion or slant the news. An honest outlier in polling can be critiqued but should not be attacked, bullied, and controversialized. As the record shows, outliers can be wrong—or correct. People can make up their own minds. The media's shaming of polls that are off trend is a relatively new concept in the age of The Narrative.

2016 Redux and Reflux

After making so many mistakes covering the 2016 campaign, we in the media promised to self-correct. But as we headed into the meaty part of Campaign 2020, it became pretty clear we were traveling the same foot-worn path.

There are two prime examples in late September 2019. First, an

NBC/*Wall Street Journal* poll generates the headline "Record 69% of voters say they dislike Trump personally," regardless of their policy views.

I take a look at the details of the poll and find other headlines that could have been chosen if one wished to spin it in a different direction. (Nobody did.) Forget about whether people say they like him personally, Trump's "Very Positive" rating in the same poll matched his record high in February 2019! What's more, his "Very Positive" ranking bested more than fourteen years of George W. Bush's "Very Positive" numbers, from July 2005 through the date of the new 2019 poll. It beat out more than nine years of Bill Clinton's numbers, from June 2001 to September 2010. And it topped Barack Obama's numbers from June 2013 through April 2016. Not only did Trump's "Very Positive" number (30 percent) equal his record high from February 2019, it was also double the number he'd had just before he was elected (15 percent).

Another relevant data point is the matter of who were the respondents in the NBC/*Wall Street Journal* poll. Thirty-eight percent said they had voted for Trump, but 47 percent had voted for someone else (mostly Hillary Clinton). In other words, the poll that purported to find Trump reaching a "record" high in terms of dislike had surveyed a sample of people who voted against Trump by a margin of 10 percentage points. Had the poll surveyed a more representative selection, his numbers would have come out differently. Had the media not been looking for a disparaging finding, they would not have hidden his gangbuster "Very Positive" numbers that had beaten out nearly three years of Barack Obama's.

On the heels of that poll, in late September 2019, a "whistleblower" said to be connected to the intelligence community makes anonymous allegations that President Trump had an inappropriate telephone call with Ukraine's president. Democrats stoke the controversy and seize upon the publicity generated to demand President Trump's impeachment—yet again. This time, Congresswoman Nancy Pelosi, the Democrats' leader in the House of Representatives, moves the ball forward with a special announcement indicating there will be an impeachment inquiry.

Next comes the inevitable polls.

A CBS News poll headlines the findings "CBS News Poll: Majority of Americans and Democrats [55 percent to 45 percent] approve of Trump impeachment inquiry." The implication of this headline is pretty clear: *Things have finally changed for ol' Trump. The scales are tipping the wrong way. Even his onetime supporters are finally turning on him! The end is nigh.*

But as usual, there is more than meets the eye. I take a look at the raw data from the CBS News poll and find one could have chosen many other headlines and statistics to give an entirely different perspective, had anyone wished to do so. For example, an accurate headline could well have been "Majority of Americans *in Democrat-Heavy Poll* Favor an Impeachment Inquiry into President Trump."

Yes, like nearly every other poll I have dug into in the past couple of years, it surveyed more Democrats than Republicans. The pollsters' reasoning as to why they rely on Democrat-heavy samples has to do with the idea that more Americans are registered to vote as Democrats than as Republicans. But pollsters acknowledge they do not actually know how this split relates to who actually turns out at the polls. Additionally, the split is not reflective of the entire population.

Digging deeper, the CBS poll interviewed 124 more Democrats than Republicans. That is a statistically significant slant toward Democrats of 6 percentage points. Let's flip the script on the sampling. Assuming, for the sake of argument, that Democrats and Republicans generally do reply along party lines, a sample that had looked at 6 percentage points more Republicans instead of Democrats would have blown the "headline" claiming "a majority of Americans" favor the impeachment inquiry. In fact, it would theoretically change the pro-impeachment majority to a minority, with 49 percent favoring the impeachment inquiry and 51 percent opposing it.

There are several other points of context that news outlets could have found worth mentioning—though few did.

In the CBS Democrat-heavy poll, a majority of respondents, 58 percent, said Trump does not deserve to be impeached or that it is "too soon to say." Adjust that for the lopsided number of Democrats interviewed, and it would theoretically become 64 percent believing that Trump doesn't deserve impeachment or that it's too soon to know.

Here's another interesting result that was not highlighted in news reports: 69 percent of Republicans who were asked about an impeachment inquiry said it makes them want to defend Trump. In other words, there is reason to believe the impeachment inquiry could be a motivational factor for Republicans in an election year—in a way that benefits Trump.

The Democrat-heavy CBS News poll also showed that 34 percent said an impeachment inquiry during the 2020 campaign would be "better for Democrats," while 30 percent said it would be "better for Trump." But had the poll interviewed 6 percentage points more Republicans than Democrats, that number theoretically could tilt in favor of Trump, 36 to 28 percent.

And last, a majority of those asked—as well as a majority each of Democrats, Republicans, and independents—said they think the primary goal of the Democrats' impeachment inquiry is to "politically damage Donald Trump's presidency and his reelection."

To summarize, a deep dive into the poll stats suggests that the following alternate headlines could have been both appropriate and accurate:

"MAJORITY OF AMERICANS IN DEMOCRAT-HEAVY POLL SAY TRUMP
    DOES NOT DESERVE TO BE IMPEACHED OR IT'S 'TOO SOON' TO SAY"
"MOST REPUBLICANS SAY IMPEACHMENT INQUIRY MAKES THEM
    WANT TO DEFEND TRUMP"
"MOST AMERICANS, INCLUDING DEMOCRATS, SAY MAIN GOAL
    OF IMPEACHMENT INQUIRY IS TO POLITICALLY DAMAGE
    DONALD TRUMP'S PRESIDENCY AND REELECTION"

But no such headlines were anywhere to be seen.

CHAPTER 10

Media vs. Media

The trend of mainstream media outlets acting as police and enforcers over other media is a shocking change to our news landscape. Reporters are now less concerned with facts and more with demanding adherence to The Narrative. They determine the position that is to be taken on issues or the facts that can be written about. They use their platform to insist that theirs is the only right and correct view. They convince their colleagues that the job of a reporter is not to be neutral or fair but to take the "correct" position. They define the parameters of the language deemed acceptable or unacceptable for the media to use when covering an issue. They punish, cajole, and threaten those who do not comply. They write "news stories" criticizing journalists who do not comply. In other words, instead of covering the news, they attack those who are off narrative and cover *that* as if it is big news. Their goal is to stop the freethinking, independent interlopers. To make it where nobody dares to go off script or disclose facts or ask questions that the media bullies want to keep hidden.

I have been hit by this phenomenon numerous times. Historically, when I've become the target of media hits, it tends to mean I am hovering over a target of my own, although I do not always know at the time what that target is! This turns out to be the case in March 2020 after the coronavirus outbreak.

When the story starts, I've just finished a Taekwondo lesson with my twenty-four-year-old daughter. We walk to our cars, and I give her a face mask in case she can use it. I'd bought it years before during the Ebola scare, and it is past the expiration date listed on the packaging, so health care workers can't accept it as a donation. At the time, masks are not yet recommended for the general public, but I figure it's better than nothing as a barrier for invisible infective droplets, so I am giving it to her. She accepts it but looks at me as if thinking, "Mom's being overly worried." At the same time, I'm working to keep my husband out of public places. There's no sense in tempting fate. He is in the higher-risk category due to his age and "chronic health issues." I know about the risk factors because I recently began compiling a list of every reported US coronavirus death so far, and the vast majority are in this high-risk category. Nobody else had compiled and published the fatalities broken down by state, patient, age, and the local health department's description of the victim's medical condition prior to death. Tracking this information has been a time-consuming task, and each time I update my website or a podcast, I explicitly point out that the numbers are changing by the minute. I tell readers and listeners to consult CDC.gov for the most updated information. As I give the statistical profile of where the deaths are occurring and among whom, I state that although the reported fatalities so far are primarily among the sick and old, it does not diminish the risk to all and the serious nature of the disease. All of this echoes the information CDC and public health officials have been distributing daily.

Anyway, while I am giving the mask to my daughter, a news alert crosses my phone. It is an article in the *New York Times*—mentioning *me*. Written by Jeremy Peters, it is titled "From Jerry Falwell Jr. to Dr. Drew: 5 Coronavirus Doubters." The subtitle reads, "While public health experts warn people to take precautions, these popular media figures insist that the virus is overhyped." According to Peters, whom I've never met or spoken with, I am one of these five popular "coronavirus doubters."

I do a triple take. I have never suggested—let alone "insisted"—that coronavirus is "overhyped." *What on earth does this article say?*

I soon discover Peters has made a shocking series of false claims in his story. He deceptively altered a quote to try to prove his narrative that I am a coronavirus doubter, and he claimed I had provided "misinformation" that put lives in danger. I quickly see that he apparently has not personally reviewed anything I have written or said about coronavirus in context. Everything he is claiming about me in his outlandish article is provably false on its face and the opposite of what I have actually said and written.

Naturally, this causes me to wonder: *What truth am I exposing that is causing somebody to try to controversialize me on this issue?*

Since the only real reporting I have done on coronavirus up to this point is compiling and publishing the comprehensive list of US coronavirus deaths, I theorize this must be what is upsetting powerful interests somewhere. Until I made my list, the main facts generally reported in the rest of the press included only the number of deaths overall or the count in a certain state. Little to no detail was given on victims' ages and previous health conditions.

No sooner is the *New York Times* hit piece published than I start getting hate comments from strangers and colleagues who have read the Peters article but not actually reviewed my articles firsthand. *Why are you putting lives in danger?* they demand to know. Some cuss me out. A former CNN colleague who had showered me with glowing remarks about my reporting over the years now contacts me on Facebook for the first time in six years to write, "I used to respect you as a straight shooting reporter who got the facts right and covered the news with integrity and you searched for the facts. Sadly, on this coronavirus [story] I think you've lost whatever credibility you had. I don't understand, because you're putting lives in jeopardy. Unfriend me if you'd like. I really don't care. I pity you."

I immediately write a footnoted letter to the *New York Times* proving that Peters had deceptively altered a quote from me, turning it into something I've never said; misrepresented a timeline; fabricated information; and failed to follow the most basic tenet taught to every first-year journalism student: to contact the subject of your story for fair comment. I demand a retraction, a correction, and an apology. The following day, I hear back from a *New York Times* editor named Carolyn Ryan. Instead of addressing the fabrications and false information, she tells me no correction or retraction is warranted because "you emphasize[d] to your listeners that the bulk of the deaths ha[d] been in a nursing home in Washington and that those who ha[d] died are [*sic*] elderly or weak or ha[d] compromised immune systems. Reasonable listeners who don't fall into one of those categories might come away with the thought that this virus poses no significant issues for them."

I am dumbfounded. The information I had reported on those matters was precisely correct and exactly what health officials were saying: the bulk of the deaths had been in a nursing home in Washington State, and most of those who had died were elderly or weak or had compromised immune systems. I cannot figure out how the *Times* could argue that reporting these facts, which were echoed frequently by public health officials and nearly all the media, including the *Times*, supposedly identifies me as a "coronavirus doubter."

Even more troubling is Ryan's outlandish implication that I should not report these facts because "Reasonable listeners who don't fall into one of those [high-risk] categories might come away with the thought that this virus poses no significant issues for them." This simple statement proves the whole point of this book. Many in the news media have redefined their role to be that of acting as censors or shapers of information in order to get the desired result in public thought. Readers and viewers, Ryan is saying, should not be told the truth because they might draw a conclusion that she and others at the *Times* apparently do not want them to draw.

As if that isn't absurd enough, I do a quick search and prove the *New York Times* had frequently reported the very same facts that its employees Peters and Ryan are now criticizing me for having reported.

"You know better," I tell Ryan in an email, with examples of the *Times*' own coverage. "The New York Times, other media and public health officials have reported the exact same facts you are criticizing me for. Under your own ridiculous standard, you have done far more than I have to make people believe coronavirus impacts the weak and elderly most. (Which is true, and that's the strangest part about your attempt at a defense of the hit job on me.)"

I then include a few lines from the *Times'* recent headlines and other coverage to punctuate the point.

How to Protect Older People from the Coronavirus
People over 60, and especially over 80, are particularly vulnerable to severe or fatal infection.

Nursing Homes Face Unique Challenge with Coronavirus
Nursing Home Hit by Coronavirus Says 70 Workers Are Sick

Of the 21 deaths across the U.S. as of Sunday, at least 16 had been linked to a Seattle-area nursing home.

A federal strike team of nurses and doctors arrived Saturday to support the staff at the long-term nursing home, Life Care Center of Kirkland, Wash., where officials have announced the deaths of 13 residents and a visitor,

Of course, it was not just me and the *Times*. Nearly everybody else was also reporting these same facts pointing to the cluster of cases and vulnerability among the aged and sick. From CNN:

Most of them were 60 years and older.

Many lived in nursing homes or other facilities.

The deadliest cluster so far has been linked to a nursing home in Kirkland, Washington. More than 20 people who lived there and someone who visited the facility have died.

People who lived in other long-term care facilities in Washington, Florida, and Kansas contracted the virus and died.

Many had other health problems,

Diabetes, emphysema, and heart problems were among the pre-existing conditions that some people suffered before they were diagnosed with coronavirus.

Why on earth is the *Times* trying to brand me a "coronavirus doubter" for reporting the same facts? Who decided to single me out and try to controversialize me?

Meantime, I'm in touch with the actor Rob Schneider, another of the five people smeared in the *Times* article. The only thing Schneider and I have in common that I know of is that we are both frequently targeted for attacks by the vaccine industry and propagandists in the media. Now here we are, lumped together in the "coronavirus doubter" article, too.

I learn that Peters deceptively manipulated a quote from Schneider in the *Times* article as he did with me. He falsely claimed that Schneider indicated, in the quote, that he had eaten at a California restaurant in defiance of a public health ban. The true, unaltered quote from Schneider clearly indicated that he had eaten out *prior to* the ban's taking effect—not after. Once the article is published, Schneider hires a libel lawyer, who sends a letter demanding a retraction. I do the same. My attorneys outline ten false statements Peters made about me in the few short sentences about me and they flag the deceptively altered quotes. The letter from my attorney to the *Times* goes on to point out:

> It comes as no surprise that my client's accurate and factual reporting [on coronavirus and who is at greatest risk] mirrors the Times' own reporting on the virus, including that, at the time of publication, the deaths in the United States numbered approximately 30 and that they were concentrated in Washington State, and a nursing home facility in particular. . . . Unlike the Times, Ms. Attkisson appropriately identified throughout her podcast that the story is evolving, and that the numbers change on a daily basis, yet the Times defamed my client for reporting—at

the same time the Times was doing so—the facts as they existed as of the broadcast . . . debunking the Times' assertion that she was downplaying the virus, or its effects, or otherwise misleading her audience regarding where and who is susceptible.

As I await a response, a *New York Times*-er who is appalled by the *Times*' accusations against me in the coronavirus article contacts me to say that Peters clearly "violated NYT code of ethics by failing to contact you and others" before publishing the story. He goes on to say that Peters's editor "was either asleep or incompetent" and "The distortions and misquoting of your work was probably beyond the capability of the line editor though they should have challenged the reporter to back up one or two things."

Considering what I've already told you about the devolution of the *New York Times*, its failure to adhere to its own standards and ethics, and the role the media now play in serving narratives, it is no surprise that we are here. A deeper look at the *Times* figures involved in this debacle reveals a lot. As for Peters, some of his colleagues say they "hold him in low regard." A fellow journalist says Peters's "reporting techniques [were already] seen as shallow, politically correct and clickbait." As for his editor, Carolyn Ryan, she was the editor in charge of the *Times*' election coverage for 2016 who—as a onetime colleague of hers notes—"Missed Trump's victory." Her reward for that big miss? A promotion at the *Times*, where she now defends shoddy journalism such as the Peters story about supposed coronavirus doubters. This is precisely what the devolution of the news has wrought.

Meantime, news coverage of the coronavirus outbreak becomes divided along political lines. For his part, Trump portrays the emergency early on as "under control" and claims Democrats are just trying to raise alarms to gain political points. Then, when it becomes clear that the crisis is not under control, he declares a "public health emergency" on January 31, 2020. He bans foreigners who recently visited China from entering the country and requires US citizens to self-quarantine upon their return. That move is criticized as "hardheaded nationalism" by writers such as *The Atlantic*'s Peter Nicholas, who writes, "Critics from WHO [the World Health Organization] and elsewhere have said the bans are unnecessary and could generate a racist backlash against Chinese people. . . . Empathy may be a casualty of Trump's own phobias: He is squeamish about contagion."

Mainstream journalists continue to provide "Exhibit A" with regard to how differently they execute their jobs today compared to a decade ago. Susan Glasser, who led left-leaning Politico's political coverage before becoming a staff writer for *The New Yorker* and a global affairs analyst for CNN, tweets:

Watching Trump TV again. POTUS [President Of The United States] attacks journalist and lies about coverage, brags about "big beautiful" wall with Mexico, veers into attack on NATO allies.

I have not yet learned anything about coronavirus response, except that Trump believes everybody is happy about it.

Jim Clancy, a retired CNN journalist, retweets the above attack on Trump and adds his own jabs, evoking a popular anti-Trump political slogan:

Resist. I am so happy I've stopped tuning in these propaganda broadcasts by @realDonaldTrump.

These tweets reveal serious tone deafness on two fronts. There was a time when few serious journalists would have expressed such blatantly slanted political opinions lest they be judged as unable to be neutral reporters of fact. But such sentiments are so commonplace today that there is not even a hint of trepidation. Reporters are proud of it. Also, the denunciation of the president comes at a time when he has just scored the highest approval of his presidency to date, with a 60 percent approval rating on his handling of the coronavirus crisis in the latest Gallup poll. Those in the media who are working so hard to tear down the president are in the minority, yet they behave with utter disregard for the idea that there are other viewpoints to consider.

At the White House briefings, some reporters ask question after question challenging Trump on his demeanor and characterizations rather than asking about matters that the public at large wishes to know about. These reporters are not seeking facts; they are looking for a "gotcha" moment that will create a buzz on social media and get kudos from colleagues, maybe land them a contributor job at CNN or MSNBC. For example, nobody asks to see the initial computer modelings or projections on which the drastic emergency measures are based. They are too busy demanding to know why Trump insists on calling it the "Chinese virus"—ignoring their own use of that very phrase in the first weeks prior to a propaganda campaign to controversialize the use of the term "Chinese." *Down the memory hole it goes.*

"Why do you insist on calling this the 'Chinese' virus?" yet another reporter asks Trump at a White House briefing. The president has already answered versions of the same question at least a dozen times.

"It comes from China," replies Trump. He adds that he is trying to dispel Communist Chinese propaganda that claimed US soldiers were part of a secret plot to unleash the virus.

Another reporter tells President Trump that someone in the administration had referred to coronavirus as "kung flu." This reporter uses up her question at the White House briefing to ask President Trump if he thinks the slur is appropriate. Trump asks her who supposedly used the phrase—and she doesn't know.

SUBSTITUTION GAME: Numerous reporters routinely attack Trump in an overtly slanted way at the White House coronavirus briefings. But when a lone reporter asks a softball question on the other side of the bias scale, her colleagues treat her as a pariah, apparently blind to the fact that she is simply a mirror image of them. Displaying anti-

Trump bias is considered fair, but displaying pro-Trump bias is treated as if it is beyond the pale.

This imbalance is demonstrated on March 19, 2020, when Chanel Rion of One America News Network asks President Trump the following question: "Major left-wing news media, even in this room, have teamed up with Chinese Communist Party narratives, and they're claiming you're racist for making these claims about 'Chinese virus.' Is it alarming that major media players, just to oppose you, are consistently siding with foreign state propaganda, Islamic radicals, and Latin gangs and cartels? And they work right here at the White House with direct access to you and your team?"

The question affords Trump the opportunity to bash Rion's liberal colleagues. "It amazes me when I read the things that I read," he replies. "It amazes me when I read the *Wall Street Journal* which is always so negative, it amazes me when I read the *New York Times,* it's not even—I barely read it. You know, we don't distribute it in the White House anymore, and the same thing with the *Washington Post*."

After the briefing, Rion pays the price for asking Trump a softball question rather than attacking him. *This is never supposed to be done.* The media-vs.-media news coverage takes off. Rion's question and Trump's response to it are attacked in stories that turn into headline news. Once again, the actual facts of an important story become secondary to the manufactured drama and media narratives. Rion tweets a photo of an anonymous typed note that a press colleague apparently left at her White House workstation. It reads, "Do you think your question was helpful in halting the spread of the coronavirus?"

Maybe not. But neither are the many questions asked by the anti-Trump media. Yet nobody leaves anonymous notes on their desks.

As the crisis drags on, coronavirus narratives become weaponized in a more blatant fashion. The liberal smear group Media Matters attacks media personality Mike Rowe as a "coronavirus doubter" in much the same way Jeremy Peters of the *New York Times* had attacked me. A well-funded, well-organized effort is clearly afoot, though I am not yet clear as to who is pulling all the strings. I only see the evidence that it exists. My name is also raised negatively in other left-leaning publications, including *Vanity Fair*, the *Washington Post*, and the Associated Press. Malicious Wikipedia editors, including one who goes by the name Philip Cross, get busy vandalizing my Wikipedia biography, as they often do. They remove accurate, footnoted material about my reporting and insert false, slanderous information from Media Matters, even though Wikipedia's official policy supposedly prohibits using partisan blogs as sources.

For his part, Rowe fights back. On his public Facebook page, he disassembles the sloppy Media Matters blog written by Madeline Peltz point by point. He points out that nothing that he has written or said "suggests that I've 'downplayed' the risks of COVID-19 in any way whatsoever." Sounds familiar.

I soon learn that the propaganda effort to attack certain reporters and personalities as "coronavirus doubters" is taking hold in the scientific world, too. As part of my reporting, I am in contact with numerous scientists who are researching coronavirus for the federal government or at academic institutions. I pick their brains, and they provide me with data, some of which is at odds with what is being widely reported. So I ask why they are not speaking out to correct the record. One by one, they tell me they worry that if they "speak like this in public," they will falsely be labeled "coronavirus doubters." In fact, none of them doubt coronavirus at all. They are very concerned. They are simply reporting factually correct information and data. They tell me that their scientific colleagues are also worried about stepping up to correct widely reported misconceptions about coronavirus for fear of "appearing to contradict Dr. Fauci or national policy." When one scientist discusses this with me at length, I comment that it's a worrisome and dangerous time when scientists and journalists alike become hesitant to report factual information or ask reasonable questions for fear of being bullied by peers and controversialized by propaganda campaigns.

It is an all-out information war. Coronavirus, a health crisis, has been weaponized and politicized in the media, each side blaming the other for hype and panic—or doubt and disinformation.

On April 3, 2020, Comedy Central's *The Daily Show* with Trevor Noah gets in on the act. As I wrote in *The Smear*, popular media personalities, from talk show hosts to comedians, are often part of the effort to promulgate smears and narratives. In this case, *The Daily Show* edits together a video montage taking potshots at conservative personalities and Trump officials, primarily those who regularly appear on Fox News. The package of brief clips from January through late March implies that these figures have missed the mark on coronavirus dangers, downplaying or denying the pandemic.

"Hannity. Rush. Dobbs. Ingraham. Pirro. Nunes. Tammy. Geraldo. Doocy. Hegseth. Schlapp. Siegel. Watters. Dr. Drew. Henry. Ainsley. Gaetz. Inhofe. Pence. Kudlow. Conway. Trump," reads a *Daily Show* tweet promoting the video. "Today, we salute the Heroes of the Pandumbic."

Liberal media and readers circulate the video, often adding remarks ranging from gleeful to hate filled. Some suggest Fox News should be sued by coronavirus victims. Among the most charged language is that used in the blog Bulwark. Its writer, Jonathan Last, attacks conservative radio host Rush Limbaugh, claiming Limbaugh has downplayed the coronavirus crisis. "This vile, foolish man has blood on his hands," Last writes. ". . . When this is all over, there should be a reckoning—a very real, very thorough reckoning—for all of the people who made this pandemic worse by pushing disinformation and lies in the service of making it harder for the country to quickly respond to the crisis."

Last's critique of Limbaugh is retweeted and amplified by the *New York Times'* Peters, who appears to have become nothing short of obsessed with writing about those he represents as conservative bad actors in the coronavirus information war. In one rambling article, he attacks "Right wing media stars" whom he says have sowed "doubt, cynicism and misinformation" using "us-vs.-them" rhetoric.

The fact is, the comments about coronavirus lampooned in the *Daily Show* video mirror similar remarks made by public health officials and liberal sources. Yet *they* escape criticism. It is comparable to the Peters article defaming me as a coronavirus doubter for making factually correct remarks that were no different than what his own newspaper had reported. A propaganda campaign seems to be afoot.

Ironically, even as Peters continues launching attacks on others in the media for alleged coronavirus misinformation, he is forced to admit errors in his "coronavirus doubters" article about me. Yes, after my attorneys contacted the *New York Times*' attorneys, the *Times* made multiple revisions to Peters's false article. Editors removed a paragraph, partially fixed a deceptively edited quote, and added a correction. It is a victory for truth, but a small one. I know how the game is played. Millions of people read the defamatory, false article; almost none of them will see the corrections that took two weeks to make.

Likewise, the *Daily Show* video is circling the earth, stoking hatred against those featured in it. Viewers are not told that many of the criticized comments were absolutely true when they were made—and that many are still true today. Unsuspecting viewers are also not told that similar comments were made by public health officials and liberal sources—a point neatly deposited down the memory hole because it undercuts The Narrative that liberals are telling the truth, focused on science, and working to keep us safe while conservatives are not.

To underscore my point, here are a few examples of remarks made by personalities criticized in the *Daily Show* video, compared to similar comments made by health authorities and left-leaning reporters.

Criticized in the *Daily Show* video:
SEAN HANNITY, FOX NEWS, FEBRUARY 27, 2020: "The sky is falling, we are all doomed, the end is near . . . or at least that's what the media mob would like you to think."

Not featured in the video:

ROBERT DINGWALL, WIRED, JANUARY 29, 2020: "We Should Deescalate the War on the Coronavirus."

NEW YORK GOVERNOR ANDREW CUOMO, MARCH 24, 2020: "It is about the vulnerable. It's not about 95% of us. It's about a few percent who are vulnerable. That's all this is about. Bring down that anxiety, bring down that fear, bring down that paranoia."

DR. DAVID L. KATZ, THE NEW YORK TIMES, MARCH 20, 2020: "Is Our Fight Against Coronavirus Worse than the Disease?": "As much as 99 percent of active cases in the general population are 'mild' and do not require specific medical treatment."

CENTERS FOR DISEASE CONTROL AND PREVENTION (CDC), APRIL 5, 2020: "The immediate risk of being exposed to this virus is still low for most Americans."

Criticized in the *Daily Show* video:

RUSH LIMBAUGH, FEBRUARY 24, 2020: "The hype of this thing as a pandemic . . . as 'Oh my God, if you get it, you're dead.'"

Not featured in the video:

NEW YORK GOVERNOR ANDREW CUOMO, MARCH 23, 2020: "Many people will get the virus, but few will be truly endangered. Hold both of those facts in your hands: many will get it, up to eighty percent may get it, but few are truly endangered and we know who they are."

REBECCA FALCONER, AXIOS, QUOTING CUOMO, MARCH 2, 2020: "The general risk remains low in New York." "There's no reason for undue anxiety."

JOHN BACON, USA TODAY, MARCH 16, 2020: "Coronavirus Not a Global Health Crisis, WHO Says."

Criticized in the *Daily Show* video:

FOX NEWS HOST PETE HEGSETH, MARCH 8, 2020: "The more I learn about coronavirus, the less concerned I am."

Not featured in the video:

ROBERT CUFFE, BBC, MARCH 24, 2020: "The UK government's chief medical adviser, Professor Chris Whitty, says even though the rates are higher for older people, 'the great majority of older people will have a mild or moderate disease.'"

VICE PRESIDENT MIKE PENCE, MARCH 4, 2020: "The risk to the American public of contracting the coronavirus remains low. To be clear: If you are a healthy American, the risk of contracting the coronavirus remains low."

DR. ANTHONY FAUCI, WHITE HOUSE CORONAVIRUS TASK FORCE, JANUARY 21, 2020: "This is not a major threat to the people in the United States and it is not something that the citizens of the United States right now should be worried about."

KATIE HAFNER, THE NEW YORK TIMES, MARCH 14, 2020: "Amid the uncertainty swirling around the coronavirus pandemic stands one incontrovertible fact: The highest rate of fatalities is among older people, particularly those with underlying medical conditions."

Criticized in the *Daily Show* video:

FOX NEWS HOST JEANINE PIRRO, MARCH 8, 2020: "It's a virus . . . like the flu . . . the talk about coronavirus being so much more deadly doesn't reflect reality."

FOX NEWS MEDICAL CONTRIBUTOR DR. MARC SIEGEL, MARCH 6, 2020: "This virus should be compared to the flu because at worst . . . worst case scenario it could be the flu."

FOX NEWS CORRESPONDENT GERALDO RIVERA, FEBRUARY 28, 2020: "The far more deadly, more lethal threat right now is not the coronavirus it's the ordinary old flu."

PRESIDENT DONALD J. TRUMP, MARCH 9, 2020: "This is like a flu."

Not featured in the video:

NEW YORK CITY HEALTH COMMISSIONER OXIRIS BARBOT, FEBRUARY 3, 2020: "We are encouraging New Yorkers to go about their everyday lives and suggest practicing everyday precautions that we do through the flu season."

ASSOCIATED PRESS, QUOTING THE WORLD HEALTH ORGANIZATION (WHO), MARCH 8, 2020: "The virus is still much less widespread than annual flu epidemics, which cause up to 5 million severe cases around the world and up to 650,000 deaths annually, according to the WHO."

ALLISON AUBREY, NPR, JANUARY 29, 2020: "Worried About Catching the New Coronavirus? In the U.S., the Flu Is a Bigger Threat."

DAN VERGANO, BUZZFEED, JANUARY 29, 2020: "Don't Worry About the Coronavirus. Worry About the Flu."

BOB HERMAN, AXIOS, JANUARY 29, 2020: "Why We Panic About Coronavirus, but Not the Flu": "If you're freaking out about coronavirus but you didn't get a flu shot, you've got it backwards."

KAISER HEALTH NEWS, JANUARY 24, 2020: "Something Far Deadlier than the Wuhan Virus Lurks Near You."

VANDERBILT UNIVERSITY PROFESSOR OF PREVENTIVE MEDICINE WILLIAM SCHAFFNER, JANUARY 24, 2020: "When we think about the relative danger of this new coronavirus and influenza … Coronavirus will be a blip on the horizon in comparison."

SOUMYA KARLAMANGLA, LOS ANGELES TIMES, JANUARY 31, 2020: "For Americans, Flu Remains a Bigger Threat than Coronavirus": "Unlike the coronavirus, which so far hasn't led to any deaths in the U.S., influenza has killed approximately 10,000 Americans since October, according to federal data released Friday." "A much deadlier killer already stalking the United States has been largely overshadowed: the flu."

UNIVERSITY OF CALIFORNIA, RIVERSIDE, EPIDEMIOLOGIST BRANDON BROWN, JANUARY 31, 2020: "Here in the U.S., [flu] is what is killing us. Why should we be afraid of something that has not killed people here in this country?" "I think we need to shift our attention back to the flu."

MICHAEL DALY, THE DAILY BEAST, FEBRUARY 6, 2020: "The Virus Killing U.S. Kids [flu] Isn't the One Dominating the Headlines."

LENNY BERNSTEIN, THE WASHINGTON POST, FEBRUARY 6, 2020: "Get a Grippe, America. The Flu Is a Much Bigger Threat than Coronavirus, for Now."

WENDY PARMET AND MICHAEL SINHA, THE WASHINGTON POST, FEBRUARY 3, 2020: "Why We Should Be Wary of an Aggressive Government Response to Coronavirus."

Criticized in the *Daily Show* video:
FOX NEWS HOST JESSE WATTERS, MARCH 3, 2020: "If I get it, I'll beat it."

Not featured in the video:
THE LANCET, MARCH 12, 2020: Death rates are lowest for those under 30; deaths are at least five times as common for people with diabetes, heart disease, high blood pressure, and other underlying diseases; the median age of patients at the time of death is seventy; and there is a low rate of infections among children.

SURGEON GENERAL JEROME ADAMS, MARCH 6, 2020: "What we want most of America to know is that you're not at high risk for getting coronavirus, and if you do get it you are likely to recover. Ninety-eight, 99 percent of people are going to fully recover."

RICARDO ALONSO-ZALDIVAR, ASSOCIATED PRESS, MARCH 21, 2020: "For most people, the coronavirus causes only mild or moderate symptoms, such as fever and cough."

DR. DAVID L. KATZ, NEW YORK TIMES, MARCH 20, 2020: "Is Our Fight Against Coronavirus Worse than the Disease?" "As much as 99 percent of active cases in the general population are 'mild' and do not require specific medical treatment. The small percentage of cases that do require such services are highly concentrated among those age 60 and older, and further so the older people are."

Criticized in the *Daily Show* video:
DR. DREW PINSKY, MARCH 2, 2020: "It's milder than we thought . . . the fatality rate is going to drop."

Not featured in the video:
KATHLEEN DOHENY, WEBMD, MARCH 31, 2020: "The fatality rate from COVID-19 is not as high as experts have reported, according to a new analysis published Monday in *The Lancet Infectious Diseases*."
DR. ANTHONY FAUCI IN THE NEW ENGLAND JOURNAL OF MEDICINE, MARCH 26, 2020: "The case fatality rate may be considerably less than 1%. This suggests that the overall clinical consequences of Covid-19 may ultimately be more akin to those of a severe seasonal influenza (which has a case fatality rate of approximately 0.1%)."
THE ATLANTIC, FEBRUARY 24, 2020: "Most cases are not life-threatening."

Criticized in the *Daily Show* video:
REPRESENTATIVE DEVIN NUNES, MARCH 15, 2020: "If you're healthy, you and your family, it's a great time to just go out, go to a local restaurant."
SENATOR JAMES INHOFE, MARCH 11, 2020: "Wanna Shake Hands?"

Not featured in the video:
DR. PETER HOTEZ, PRIOR TO MARCH 27, 2020: "Historically travel bans tend not to work very well, they tend to be counterproductive."
ROSIE SPINKS, THE NEW YORK TIMES, FEBRUARY 5, 2020: "Who Says It's Not Safe to Travel to China? The coronavirus travel ban is unjust and doesn't work anyway" "The coronavirus outbreak seems defined by two opposing forces: the astonishing efficiency with which the travel industry connects the world and a political moment dominated by xenophobic rhetoric and the building of walls."
NEW YORK CITY MAYOR BILL DE BLASIO, MARCH 2, 2020: "I'm encouraging New Yorkers to go on with your lives + get out on the town despite Coronavirus."
NEW YORK CITY HEALTH COMMISSIONER OXIRIS BARBOT, JANUARY 27, 2020: People "who had recently traveled from Wuhan were not being urged to self-quarantine or avoid large public gatherings." "There is no reason not to take the subway, not to take the bus, not to go out to your favorite restaurant, and certainly not to miss the parade next Sunday."

"As we gear up to celebrate the #LunarNewYear [Chinatown parade] in NYC, I want to assure New Yorkers that there is no reason for anyone to change their holiday plans, avoid the subway, or certain parts of the city because of #coronavirus. . . . We are here today to urge all New Yorkers to continue to live their lives as usual." "There's no risk at this point in time . . . about having it be transmitted in casual contact, right?" "The risk for New Yorkers of the coronavirus is low, and our preparedness as a city is very high."
HOUSE OF REPRESENTATIVES SPEAKER NANCY PELOSI, FEBRUARY 24, 2020: Urged people to visit San Francisco's Chinatown. "That's what we're trying to do today is to say everything is fine here. Come because precautions have been taken. The city is on top of the situation."
MAYOR OF FLORENCE, ITALY, DARIO NARDELLA, FEBRUARY 2, 2020: Suggested that city residents hug Chinese people to encourage them in the fight against the novel coronavirus.

Again, it is important to emphasize that I am not saying conservatives and liberals were equally as wrong. Some of these highlighted statements—though criticized—proved absolutely correct! The point is that similar statements made by Democrats, Republicans, and health experts were treated differently depending on whom the media wished to controversialize and The Narrative they wished to forward.

Interestingly (but not surprisingly if you've made it this far in the book), many of the quotes that were singled out for criticism by the left had been written up by the propaganda group Media Matters. The media and Comedy Central took their signals, if not marching orders, from the partisan smear group.

With the *Daily Show* video and other attacks made against select people only, while others are exempted after making the very same comments, the clear message being delivered to all is: *We are turning this into a battle of Right vs. Left and using it as a tool to undermine those we want to controversialize. When you do not depict coronavirus as the worst-case scenario, we may accuse you of being conservative, clueless, and anti-science. Therefore, you must self-censor the truth or lie, rather than present accurate information to the public. If you don't obey, we'll destroy you as a "coronavirus denier" even though you are reporting factually and responsibly and not "denying" anything.*

It may be inevitable in today's politically dominated news media environment that coverage of an international health crisis would get divided up into camps of Left vs. Right, liberal vs. conservative, pro-Trump vs. anti-Trump. Considering all we have discussed, it should shock no one that hidden interests are trying to shape the parameters of what can and must not be said on an important topic. But with such glaring inconsistencies in the media's treatment of people reporting the very same information, public trust in the media continues to be eroded. And few minds are changed by the propagandists.

But the bigger lesson is how much media bandwidth has been taken over by this media-vs.-media mentality. Instead of publishing articles with information on topics of importance to the general public, the media are filling space, taking up time, and assigning reporters to attack other reporters, stories, and media figures.

What could we be learning if all of this effort were devoted to covering actual news?

What are we missing?

CHAPTER 11

Media Mistakes

There are countless lists of President Trump's "lies," particularly among left-leaning media organizations. The *New York Times* counts them. Politifact by the Poynter Institute keeps a running tally. The *Washington Post* counted 16,241 lies or misleading statements in Trump's first three years. CNN tracks them regularly. I am not sure how much is added to the information landscape by having so many news operations spending so much time and resources reporting on the very same thing. But they're obviously trying to make a point.

Meantime, I noticed that we in the media were committing a whole lot of our own mistakes, yet nobody was making comprehensive lists of our misreporting. I see it as the flip side of the same coin: as we accuse Trump of unprecedented dishonesty, we seem utterly blind to the error of our own ways. Does any politician's lie excuse ours? Can we really hide behind claims that when he makes misstatements, they are somehow worse than ours?

Once President Trump was elected, it quickly became painfully clear that many in the media were so out to get him that they became shamefully sloppy. Since I saw nobody else compiling a list of their errors, I began tracking major media mistakes in the era of Trump. From the time of his election in 2016 through July 13, 2020, I counted 131 of them. When the same mistake was made by numerous news organizations, I counted it as only one. Here is an example.

On May 29, 2019, The Narrative is in full force. The *Wall Street Journal* and other news organizations break a story that exploits bad blood between President Trump and the late Republican senator John McCain of Arizona.

The story states or implies the following key points:

CLAIM ONE: The Navy hung a tarp over the name of the Navy destroyer USS *John S. McCain* so that President Trump wouldn't see it during his visit to Yokosuka, Japan.

CLAIM TWO: A barge was placed to block Trump's view of the ship, lest the McCain name offend or anger him.

CLAIM THREE: Sailors from USS *McCain* who wore hats bearing the ship's name were turned away from the presidential event or given the day off, so that Trump would not see the McCain name.

CLAIM FOUR: Acting Defense Secretary Patrick Shanahan knew of the above-mentioned "scheme" to make sure the McCain name did not cross Trump's view.

News stories attribute these McCain name-hiding maneuvers to "the White House," which many readers infer to mean President Trump himself gave the orders. In other words, *Trump is such a baby, so self-absorbed, so jealous, so unconcerned with the stature of his job, and so triggered by the name of a political rival that he would order his staff to stage ridiculous modifications at a military appearance to accommodate his outrageous whims.*

Even if true, the news on its face would be of little importance in a neutral news environment. A few years ago, with rare exceptions, if such information were confirmed, it might merit a brief mention on the national news and then persist only in the form of newsroom cackles and gossip blogs. But in 2019, the story is treated as if it were worthy of international attention. It is amplified far beyond its relative significance, with headlines crafted to reinforce anti-Trump narratives.

By way of background, if you aren't familiar with it, the nasty feud between Trump and Senator McCain began in July 2015. McCain referred to Trump supporters as "crazies." Trump counterpunched by deriding McCain's Vietnam War service and prisoner-of-war status. Adding to the quarrel, in 2016, McCain secretly passed along the infamous "Steele dossier" anti-Trump opposition research to the FBI. And McCain's aide allegedly leaked it to the press to try to damage Trump. McCain also cast the deciding vote to kill Trump's planned repeal of Obamacare in a dramatic flourish on the Senate floor late at night on July 28, 2017. Trump frequently criticized McCain for that vote thereafter. And of course, Trump needled McCain by pointing out that he—Trump—had won the White House whereas McCain had failed. One could say the two men trolled one another like schoolyard kids. But the prevailing media story in every case is that Trump is petty; McCain is a hero.

The May 2019 news story about USS *McCain* dominates the bandwidth on major national news outlets for days on end. Reporters quote McCain's daughter Meghan, who tweets:

> Trump is a child who will always be deeply threatened by the greatness of my dads [*sic*] incredible life. There is a lot of criticism of how much I speak about my dad, but nine months since he passed, Trump won't let him RIP. So I have to stand up for him.
>
> It makes my grief unbearable.

But shortly after the initial news reports, something happens. Key parts of the anti-Trump story line begin to fall apart.

It appears to be true that, in advance of Trump's trip to Yokosuka, an unidentified US military official had sent an email directing that USS *McCain* be kept from Trump's view. However, it also appears *that direction was never followed*. Which pretty much changes the whole narrative.

The claim that a tarp was placed to block the ship's name from Trump's view was false, according to US military officials. In fact, they say, a tarp that had covered part of the ship during maintenance prior to Trump's visit was *removed* for Trump's visit.

"The tarpaulin was used as part of hull preservation work on the McCain and was removed on Saturday, two days before Trump delivered a Memorial Day address at US Naval Base Yokosuka, where the McCain was stationed," says Commander Nate Christensen, a spokesman for the US Pacific Fleet, in a statement after the news story goes global.

Additionally, US officials say that a paint barge anchored in front of USS *John S. McCain* was moved out for Trump's visit and gone by the time the president arrived.

"All ships remained in normal configuration during POTUS' [the President's] visit," says Commander Christensen.

With major components of the *Wall Street Journal* story called into serious question, the *New York Times'* Maggie Haberman fires off an odd tweet. She ignores the contradictions and states that her own newspaper has confirmed the *Wall Street Journal*'s "excellent scoop," including the claim that the sailors wearing USS *John S. McCain* hats had been turned away from the Trump speech. This demonstrates how when doubts are raised about the truth of a narrative, those who support it tend to double down rather than correct their mistakes. For her part, Haberman is well known in the business of narrative pushing. Hillary Clinton campaign strategists considered her a go-to when they wanted to plant stories in the press. That much was revealed by internal documents hacked from the Democratic National Committee in 2016. Emails dated January 2015, when Haberman worked for Politico, show Clinton officials planning to use Haberman as a tool to put out some of their propaganda under the guise of a news story. They referred to her as a "friendly journalist" who had "teed up" stories for them in the past and "never disappointed" them. Now promoted to the *New York Times*, Haberman covers Trump, usually generating or furthering the anti-Trump narratives *du jour*.

As the story falls apart with each passing hour, CBS News actually goes against The Narrative and weighs in with context that casts further doubt on the initial reporting. CBS News Pentagon correspondent David Martin explains why USS *McCain* ball cap–wearing sailors might really have been turned away from the Trump event: not because of the McCain name the hats bore but because of the dress code. He points out that eighty sailors from more than twenty ships and Navy commands were present for the president's visit, and "all wore the same Navy hat that has no logo, rather than wearing individual ship or command hats. . . . it is possible the reason they were turned away is that ball caps were not part of the dress code for the event."

In his reporting, CBS's Martin isn't defending Trump; far from it. He is simply doing what news reporters who aren't pushing narratives do: providing both sides of the story, including context and counterpoints.

As for the reporting claiming Acting Defense Secretary Shanahan supposedly knew of the "scheme" to keep the McCain name from Trump's view, that appears to be yet another error. Shanahan firmly denies that he knew of any such thing. President Trump and White House aides tell the press that Trump had no knowledge of it, either.

The initial story bore all the hallmarks of a narrative rather than fact-based news reporting. It relied on unconfirmed information. It was reported in a slanted fashion. It was amplified in the media beyond its organic importance. Little to no regard was given to potential counterpoints.

At this point, journalists and commentators who got the facts wrong—or at least told disputed and incomplete stories—might at least consider quietly moving on to another subject. Instead, they twist themselves into knots, adjusting and morphing their reporting to incorporate the corrected information while pretending their original stories were accurate. They are collectively watching each other's backs, bitterly clinging to The Narrative rather than admit they got something wrong.

Another example is a commentary by Tiana Lowe in the *Washington Examiner*. After President Trump and others criticize the mistaken reporting, Lowe concedes a major point. She acknowledges that Defense Secretary Shanahan may not have known about the "scheme" to keep USS *McCain* from Trump's view, after all. But then she attempts to convince us that the rest is true. She asks, "Save for the Shanahan allegation, what part of this story is fake?"

With a single figurative stroke of a pen, Lowe acknowledges fault with the story even as she implies that there is nothing to fault. She continues in an incredible feat of hair splitting, "Trump denies ordering the plan, but the journalists who wrote these stories never accused him of doing so." (Of course, that's just what the journalists implied, and without that implication, it's hard to imagine how it would even have been a news story.)

"The plan to obscure the ship clearly wasn't executed by the time Trump arrived in Japan, let alone at Yokosuka," Lowe admits in what is actually a major correction, "but the journalists who wrote these stories never said it was." (That implication was the whole point of the story.)

Journalistic ethics dictate that the errant reporters should have issued a mea culpa along the lines of this:

Key points of our original story have come under dispute and proven incomplete, inaccurate, or lacking in important context. As for the claim that Acting Defense Secretary Patrick Shanahan knew of a supposed "scheme" to make sure the McCain name did not cross Trump's view during the president's visit to Japan, there is no evidence of a scheme and Shanahan denies knowing of any such thing. Further, none of the supposed actions to block Trump from seeing "McCain" actually took place. As for the claim that the Navy placed a "tarp" to cover the name of the USS McCain: that is untrue. No tarp covered the ship while President Trump was in Japan. In fact, a tarp that had covered part of the ship during maintenance was removed for Trump's visit. As for the claim that a barge was placed to block Trump's view of the USS McCain: that was also untrue. All ships remained in their normal configuration for the president's visit. There is also no evidence to support the implication that sailors wearing "USS McCain" ball caps were turned away so that Trump would not see the name of his political rival. The ball caps may have simply violated the applicable dress code since all admitted sailors wore Navy hats bearing no individual ship logos.

Not surprisingly, no such message was delivered. Instead, the media message, if you read between the lines, was something along the lines of this:

Our story was one-sided and incomplete, but it's still accurate. We didn't exactly state what we implied, so we weren't really wrong. You see how Trump twists our words! This all proves our point about Trump more than ever!

Days later, a false headline on the left-wing entertainment news website Deadline still remained: "[McCain] Family Name Obscured on Ship to Prevent Trump Trigger in Tokyo." Nearly all now agree the McCain family name was never "obscured" as the headline states. But The Narrative survives the truth—and that's the objective.

The McCain story is just one example of the media's anti-Trump-narrative-at-any-cost behavior. It's a pattern. Never before have so many formerly well respected national news outlets made so many major reporting errors. Taken individually, each error could theoretically be the result of honest mistakes or sloppy reporting. Collectively, they carry the hallmarks of a Narrative. First: The mistakes all seem to go in one direction—against Donald Trump. Second: They are the kinds of mistakes that novice journalism students know better than to make, but they are being made by seasoned journalists at top news outlets. Third: There are rarely any significant consequences for the journalists who make the mistakes, corrections are few and far between, and apologies are almost nonexistent. Instead, we excuse the errors and indignantly declare, *"We corrected our mistake as soon as we learned we were wrong. That proves just how credible we are!"*

Why aren't reporters fired for inaccuracies when the whole job of reporting rests on accuracy? Why do news reporters continue to consult the same news sources after those sources get caught providing false information? All of this makes sense only when you understand that the goal isn't to shed light on facts, information, and truth; it's to further The Narrative. Looking through that lens, it's easy to see that even when information is wrong, if it advances The Narrative, the mission is accomplished.

Media Mistake Number 101: A Doozy

On Thanksgiving Day 2019, we passed the one hundredth major media mistake in the era of Trump, by my count. This particular error is worth dissecting because it is reflective of the bias and sloppiness endemic in so many other media mistakes.

This error was committed by *Newsweek*'s political reporter Jessica Kwong, who wrote a story and published a tweet asking "How did Trump spend Thanksgiving? Tweeting, golfing and more." She went on to write that Trump was spending Thanksgiving Day at his Mar-a-Lago resort in Palm Beach, Florida. The story implied Trump was once again goofing off, compared to his heroic predecessor, President Obama, who used to do only selfless things.

The problem is: her story was false.

Trump had actually left Florida the evening before Thanksgiving to fly to Afghanistan, where he spent the holiday not golfing but serving dinner to US troops. When her mistake became clear, Kwong claimed she'd made an "honest mistake." On Twitter, she wrote that she was deleting her earlier, incorrect tweet about Trump because "it was written before knowing about the president's surprise visit to Afghanistan."

There are five key problems from an ethical and journalistic standpoint. First, Kwong demonstrated a shocking lack of basic reportorial knowledge. I'm not a political reporter, nor do I closely follow the White House. Yet I knew enough to wonder whether Trump might visit the troops on Thanksgiving. All recent presidents have made holiday visits to thank our soldiers at some point. One would think a national political reporter and her editors would know to watch for a possible "surprise visit" by the president.

Second, the *Newsweek* article demonstrated an inexcusable failure to attribute. Kwong's mistake would not have been as problematic had she attributed the claim that Trump was golfing on Thanksgiving to an actual source instead of reporting it as if it were her own first-hand, confirmed information. Kwong isn't the only one who seems to have abandoned the basic journalistic practice of attributing information to its source. Reporters routinely declare information to be fact as if they have personally confirmed it, when they could not possibly have done so.

Third, there was a baffling failure to fact-check. This is one of the most basic tenets of journalism: no matter how obvious something seems, no matter how many others are reporting the same thing, no matter what a video clip seems to show—it often proves to be wrong. That is why it is so critical for reporters to check their assumptions. Kwong should have contacted the White House to see if her claim that Trump would be golfing on Thanksgiving was true. If the people there had said yes and she attributed that answer to the White House, then even if Trump ended up in Afghanistan, Kwong could not have been journalistically faulted for the mistaken information. She would have done her job correctly. Instead, she reported false information as if she had checked it.

Fourth, there was a glaring failure to correct the mistake after the fact. Although *Newsweek* fixed the story, it didn't *really* correct it. Editors called the revision an "update." This is disingenuous. The story hadn't changed or been "updated." *Newsweek* simply learned that its original report was false. That merits both a correction and an apology.

And fifth, the false information persisted well after the "update." *Newsweek* retained its false headline stating that Trump golfed on Thanksgiving. He didn't.

I think the most remarkable thing of all is that this is a case of history repeating itself. It is an example of lessons not learned by the media. NBC had generated a similar scandal less than a year before, also tying itself into a pretzel to make it seem as if it had not bungled the story.

The NBC saga began on December 25, 2018, nearly eight hours before Christmas Day's official end. The network published a headline blaring "Trump becomes first president since 2002 not to visit troops at Christmastime." The story claimed Trump had broken "from a recent tradition of actually visiting troops and wounded warriors." The article took multiple jabs at Trump to advance the anti-Trump narrative of a president who could not live up to the standards of his predecessors.

What NBC did not know was that contrary to its reporting, Trump had left the White House late on December 25, 2018, to visit US troops in Iraq. What happened when that mistake was revealed is quite telling. Like *Newsweek*, NBC was unwilling to admit its mistake. Instead of a simple apology saying it originally had no idea the president had sneaked off to Iraq and that it had made an assumption without bothering to verify it, NBC publishes a lengthy editor's note parsing the definition of what constitutes a "Christmastime visit" and claimed that the original article was *technically* correct. *Depends on what the meaning of the word "is" is.* The NBC editor makes the argument that, *okay, maybe Trump visited the troops around Christmas after all, and, yes, he did leave the White House on Christmas to head to Iraq. But he did not arrive in Iraq until after the stroke of midnight—and the day after Christmas isn't really a Christmastime visit, is it?*

So instead of an apology, NBC defends the mistake by arguing when Christmastime technically begins and ends. The NBC editor's note reads as follows:

> *As of the end of Christmas Day 2018, Trump had not visited troops during the holiday season, and had announced no plans to do so. The article was correct, but on Dec. 26, the situation changed. Trump and*

the first lady, Melania Trump, made an unannounced visit to troops in Iraq. As a result, the thrust of this article is no longer correct, even if it was at the time. In the interest of transparency, we are keeping the article on NBCNews.com so that the record will reflect the situation on the day the article was published, and are directing readers to the article about Trump's Iraq visit here. We are also altering one line in the article, as well as the headline, to be more specific and to note that Trump was the first president since 2002 who didn't visit military personnel on or before Christmas, rather than at Christmastime.

Claiming the article was "correct" "at the time" but "the situation changed," without acknowledging the journalistic malpractice at issue, is remarkable! The "corrected" article then goes on to double down on its mistake in a way that is still designed to make Trump look bad. First, NBC chooses a new tack that fits The Narrative it set out to prove and stuck by it regardless of the facts: "By staying home on Tuesday, Trump became the first president since 2002 who didn't visit military personnel *on or before Christmas* [emphasis added]."
The article then elaborates:

Based on a check of NBC logs, President Barack Obama visited troops at Marine Corps Base Hawaii, in Kaneohe Bay, every Christmas he was in office, from 2009 to 2016.

Before him, according to a check of news releases, President George W. Bush visited wounded warriors at Walter Reed from 2003 to 2008. . . .

[Trump previously said] he hasn't visited a combat zone because he's "had an unbelievably busy schedule."

Among the anti-Trump press, the new version of the story, comparing Trump unfavorably to Obama and Bush, goes viral. No attention is paid to NBC's mistake. It's like it never even happened. The focus becomes the newly revised scandal that Trump is the first president not to visit the troops *on or before Christmas Day*. One exception to the groupthink is the normally anti-Trump Erik Wemple of the *Washington Post*. He takes issue with NBC, pointing out:

> As the [press] pool report notes . . . Trump didn't stay home [Christmas] Tuesday [as NBC claimed in the updated version of its article], at least not all of Tuesday. He left the White House "late on December 25." . . .
>
> The [NBC] story appears to rest on a lawyerly definition of "Christmastime." . . . "Christmas day and the days and weeks before it."
>
> However, other common definitions of Christmastime expand that understanding to "the Christmas season, or the period from about December 24 to January 1 or January 6." But NBC seemed intent on making the story be about whether they could technically defend that they hadn't jumped the gun and made a false assumption; rather than what Trump actually did over the Christmas holiday. . . .
>
> Give the guy the credit he deserves for getting off his butt, as does The Post: "The unannounced visit continues a holiday tradition followed by past presidents."
>
> Correct the piece, NBC News, or prepare to stand legitimately accused of propagating fake news.

NBC could have chosen numerous facts to highlight about President Trump if it were focused on disseminating information instead of generating a smear. Let's look at the actual stats. It turns out President Obama never visited US troops in foreign countries or combat zones on Thanksgiving or at Christmastime, as Trump had done twice by early 2020. The closest President Obama got to spending Thanksgiving or Christmas in a combat zone with the troops was a visit to South Korea a week before Thanksgiving in 2009. President Bill Clinton was close to a Christmas visit with a December 22, 2007, visit to Bosnia after the war. The annual Christmas troop visits that NBC rushed to credit Obama for—all of those had taken place, not in a combat zone but stateside, conveniently near Obama's home in Hawaii. According to the *Los Angeles Times*, every Christmas he was in office, "Obama made the short trip from his rental home in Kailua to the Marine Corps Base Hawaii."

SUBSTITUTION GAME: This isn't to suggest that it's not meaningful that Obama spent part of his Christmases with our uniformed men and women, even if it was close to home. But if the tables were turned and Obama had been the one who traveled to a foreign combat zone over Christmas while Trump had visited only troops close to his Mar-a-Lago resort in Florida, I think you can guess what the headlines would have read.

Another point made in the original slanted NBC article about Trump was that he had "yet to visit an active combat zone," as if it were an anomaly that he had supposedly gone almost two years without doing so. In fact, according to military records, presidential visits to combat zones are relatively rare. Presidents have visited a combat zone only twenty-seven times in US history prior to Trump. (By early 2020, Trump had added at least four to that historic tally: visits to South Korea in November 2017, Iraq in December 2018, South Korea again in June 2019, and Afghanistan in November 2019—not to mention he was sometimes accompanied by First Lady Melania, who has made at least two visits to a combat zone.)

I discovered another interesting point while researching presidential trips to combat zones. Although you might not immediately think of South Korea as a combat zone, it is one. That is because our war with North Korea was never declared officially over. Both North Korea and the United States (which is allied with South Korea) patrol a demilitarized zone that divides the two Koreas. Remarkably, when counting Obama's combat zone visits, the military and media included his trips to South Korea as notches on his belt. But they did not count visits to the same place for Trump. At the very time when the press was claiming Trump had never visited a combat zone, he had made two trips to South Korea, one in November 2017 and another in June 2019.

Punctuating the point, Trump's Wikipedia page claims Trump's "first" combat zone trip was on December 26, 2018, to Iraq. It, too, ignored his earlier visit to South Korea. In contrast, the official tally for Obama counts all three of his trips to South Korea as visits to a combat zone. Apparently, among slanted media, South Korea is a combat zone when Obama visits but not when Trump does.

Here are seven other true headlines NBC could have written about Trump's Christmastime visit to Iraq—but didn't.

- "Trump Becomes the First US President in 16 Years to Visit US Troops in a Combat Zone" (since Bush in 2003).
- "Trump Is One of Only Three US Presidents to Spend Thanksgiving Day with US Troops in a Combat Zone." (The other two were President George H. W. Bush and President George W. Bush.)
- "Melania Trump Becomes the First First Lady to Spend Christmastime Visiting US Troops in a War Zone." (Thanksgiving, too!)
- "Trump Is First President Since George W. Bush to Spend Thanksgiving with the Troops in a Foreign Country."
- "President Trump Is Only the Third US President—All of Them Republicans—to Spend Thanksgiving with US Troops in a Combat Zone or US Troops Anywhere." (President George H. W. Bush spent Thanksgiving in Iraq in 1990. His son President George W. Bush spent Thanksgiving in Iraq in 2003. Obama didn't do it. Clinton didn't do it.)
- "Trump Is the Only US President to Visit the Troops in a Combat Zone So Close to Christmas Day."
- "President Trump Is the Only US President in History to Visit US Troops Both in a Combat Zone on Thanksgiving *and* So Close to Christmas."

Fast-forward to the similarly false report by *Newsweek* that claimed Trump was golfing at Mar-a-Lago on Thanksgiving Day in 2019, when in fact he was with US troops in Afghanistan. *Newsweek* got so much blowback from readers that it quickly fired the reporter, Kwong. But I was surprised to hear some journalists coming to her defense.

Several appeared on a cable television panel discussion making excuses such as "Well, she wrote the story in advance, how could she
know Trump would, surprisingly, go to Afghanistan?" And "As soon
as *Newsweek* found out, they took down the erroneous article. Honest
mistake." One journalist even likened Kwong's error to the benign
practice of journalists preparing obituaries of well-known older
people before they die, to have the obituaries ready to publish as
soon as a death is announced. "It's just that the person didn't die, in
this case," the reporter making the comparison reasoned. "Honest
mistake."

I don't think it's a valid comparison. Reporters do prepare obituaries in advance, but they write a factual look back at the notable
person's real life. In Kwong's case, she wrote a story predicting,
fabricating, or assuming what she thought would happen in the
future—without saying that was what she had done. And of course
she assumed the worst about Trump. Since she was predicting or
making up facts, she could just as well have pre-written that Trump
was spending Thanksgiving visiting the troops rather than golfing.
But the media's mistakes always seem to slant one way: toward the
anti-Trump narrative.

At least one social media observer noted that prior to getting fired
by *Newsweek*, Kwong had gotten caught up in another anti-Trump
snafu. She'd published a tweet about President Trump implying he
was stingy. Her post was prompted by a tweet by Trump adviser Har-
lan Hill showing the president giving a cash tip to the staff at Trump's
Washington, DC, hotel after dinner. Hill's tweet read:

President @realDonaldTrump just gave the staff at @TrumpDC a wad of
cash in appreciation for their great service! So generous!

Kwong's tweet in response mocked Trump's tip:

VIDEO SHOWS TRUMP GIVING CASH TO HIS D.C. HOTEL STAFF, HIS
ADVISOR BOARD MEMBER CALLS IT "SO GENEROUS."

Not all that long ago, no legitimate reporter would have considered making fun of Trump's tips to be fodder for a news headline. And any reporter on that track would have contacted the White House for comment, as routine reporting ethics require. Kwong failed to do that. So it wasn't until after her story was published that Trump's adviser had the chance to push back on her sarcasm. He explained that the cash Kwong referred to as "thin" was actually a stack of hundred-dollar bills and was in addition to a tip he had already given to the waiter. *Newsweek* adjusted its snarky headline, removing the reference to a "thin" stack of cash.

"Stack appeared thin," Kwong explained, revealing that she had based her conclusions on unverified observations of a social media video.

When reporters get caught doing bad journalism, they and their colleagues often retort that *"Trump lies more than we do."* Or *"Trump mocks the media and his enemies on Twitter."* The fact is, Trump mercilessly mocks—and is mercilessly mocked by—his political detractors such as leading congressional Democrats Nancy Pelosi, Chuck Schumer, and Adam Schiff. Collectively, political figures have no obligation to be neutral or fair. In fact, the nature of politics breeds this type of dynamic.

But journalists are different. Our professional obligation is to cover our subjects fairly. We must maintain the same high standards even when we don't like somebody we are covering—especially, perhaps, when we don't like that person. Otherwise, why have standards at all? The implication that somehow the media's mistakes and attacks are justified because of what we believe to be the flawed moral character of the target is a precarious one.

The way we cover those we perceive as the enemies of our own viewpoints defines how well we do our job covering the news.

CHAPTER 12

There's Hope

Not all good journalism wins prizes. In fact, good journalism that is counter to prevailing narratives is likely to fall under attack.

If you're old enough to know the comedian Rodney Dangerfield's schtick, "It's not easy bein' me," you have some idea of what it feels like to be a news reporter who reports on stories that powerful interests want to shape, controversialize, or hide. *It's not easy bein' them.* There's nothing for them to personally gain by independently reporting material that so many want to suppress. Their colleagues may ostracize or dismiss them. Special interests may smear them as "tinfoil hat conspiracy theorists." Organized astroturfing social media mobs may be dispatched to motivate real-life people into action against the reporter and his story. Not to mention the news bosses who would rather not deal with the phone calls and pressure coming from powerful people who object to the reporting. What boss wants that kind of a headache? If you're that kind of reporter, it becomes increasingly difficult to hang on to your job and pay the bills. But of this I am sure: there is more need than ever for that kind of reporter. Americans are thirsty for information they can trust.

Mark Levin says the modern mass media are simply not like the image of journalism that lives in our imaginations. Levin is a conservative commentator and former chief of staff for President Ronald Reagan's attorney general Edwin Meese. Levin argues the modern mass media do not favor a free press. In his book *Unfreedom of the Press,* he writes, "The American free press has degenerated into a standardless profession, not through government oppression or suppression, but through self-censorship, groupthink, bias, omission, and propaganda."

I ask Levin to elaborate when I meet him in person to talk about his research. "There's a new doctrine that's being pushed in journalism school, has been for about thirty years . . . to push what's called public journalism or community journalism, which is social activism," he tells me. "Our news is filled with phony events and filled with propaganda."

According to Levin, young journalists are taught to think of themselves as activists. So it is only logical they would believe it's perfectly natural to shape their reporting with narratives to convince the consuming public to think the "right" way. There is a long history of press pushing narratives. Levin points to America's early pamphleteers, such as Thomas Paine, who wrote to advocate for independence from Great Britain. But their narratives had a different purpose. "They wanted to fundamentally transform government, throw off the monarchy and create a representative government," he notes. "The media today want to do the opposite. They want to fundamentally transform the civil society in defense of an all powerful centralized government."

The pamphleteers admitted they were biased. "They didn't believe in objective news. They supported a cause. They didn't view themselves as seeking what's the news of the day, the information of the day. They were revolting against a tyranny," says Levin. The difference today is "the press poses as seeking objective truth when it's not."

I ask Levin whether there is any truth to the claim that Donald Trump is to blame for the news media's declining credibility. That he has been uniquely threatening to a free press.

"There's really no truth to that," says Levin. He then recites a list of US presidents who, he says, really did endanger the media. He starts with John Adams and the Sedition Act of 1798: "He imprisoned journalists, shut down some newspapers." Levin tells me that Abraham Lincoln's secretary of war Edwin Stanton "shut down newspapers, imprisoned journalists." Woodrow Wilson instituted a new Sedition Act in 1918 and "shut down newspapers, imprisoned journalists, imprisoned political opponents." Franklin D. Roosevelt "unleashed the IRS on several publishers, including Moe Annenberg, who owned the *Philadelphia Inquirer*, because they didn't like the New Deal." Barack Obama and his FBI "went after the *New York Times*, Fox, and AP" (and Sharyl Attkisson, Levin later notes) with secret subpoenas or surveillance. "Donald Trump may call the press, or a particular segment of the press, 'the enemy of the people.' But he hasn't done anything like *that*," says Levin.

As for how we might fix things, Levin advises, "First, you really should strive, if you're a newsroom, to separate news from opinion. Stop hiring ideologues in the newsroom because it's harder and harder for ideologues to be objective or as objective as they can be. Number two, maybe you're liberal, maybe you're conservative. But at least apply some objective standards and process to the gathering of news. We're not doing either now."

As grim as the situation can seem—and I know I've painted a pretty grim picture—I also try to remind people that great journalists are still doing admirable work every day. There are dedicated reporters who are taking on the powerful at their own peril. They are braving the smears and brushing aside their own self-interest to expose uncomfortable truths. You just have to figure out where to find them, and that is no easy task. Years ago, we had several trusted news sources to turn to. Maybe we watched a certain network's newscast and read a particular newspaper. But today, news consumption is hunt-and-peck style. Your favorite reporter may not work for just one news outlet; she might contribute to numerous print and video organizations. You may come to trust one publication when it comes to a particular topic of interest but look elsewhere on a different subject. Nobody reads a single newspaper "from cover to cover" anymore.

Many people are skipping traditional news outlets altogether. They get their news from a mix of quasi-news reporters, bloggers, partisans, and citizen journalists. There are several reasons for this.

First, today's broadcast, cable, and national print outlets often avoid reporting on stories, facts, and views that fight dominant narratives. Second, when these news sources do tackle an off-narrative story of interest to the public, they often end up transforming it into a narrative to spoon-feed to the public. Third, news consumers have decided that if they're going to get a narrative anyway, they may as well shop around for the one they prefer. The result is that a lot of important stories that would have been broken by CBS or the *New York Times* a few years ago are now being unearthed by an eclectic mix of those under the category of "other." They are filling a gap left by the "mainstream media."

Many of the most popular alternative information sources are politically right of center. That is because prevalent news narratives tend to gravitate toward urban, liberal interests. So a vacuum exists when it comes to news that appeals to conservatives, news that is nonpolitical, and news that is of interest to nonpartisans.

Another factor accounting for the rise in nontraditional reporting and reporters is the fact that the rest of the media seem to cover two or three stories over and over, hour after hour, day in and day out. There is little diversity or originality among them. They leave thousands of stories untouched. A colleague recently remarked to me, "I look at cable news, then I read the *Washington Post,* maybe a blog . . . not only are they all reporting the same stories, they're using the same words, citing the same sources. After reading one or two, I already know the next one is going to say the exact same thing."

The Internet is the great equalizer that has made it possible for everyone from bloggers to "citizen journalists" to step in, make videos, and act as reporters. News that would otherwise be ignored is getting covered. And that is generally a good thing. However, the nature of the nontraditional news messengers makes it all the easier for traditional media to dismiss them. The traditional media want the public to stay focused on *their* narratives. They declare their own reporting to be accurate and fair, then label those who are covering other angles as unreliable or partisan.

Sometimes the critics have a point. If you think journalism is dead within mainstream news, consider that nontraditional reporters have no professional journalism obligations at all. They have no ethical responsibility to tell a story objectively, as journalists were once expected to do. Citizen journalists, bloggers, and advocates are presenting information in the light that best suits their purposes. At times, their information may be untrue, out of context, slanted, poorly researched, or unfair.

SUBSTITUTION GAME: In reality, alternate media types are criticized for doing much the same thing that many mainstream reporters are doing: presenting stories and facts in a way that advances their views and interests. But the mainstream reporters seem to lack self-awareness. They report in a biased fashion and declare it "fair." But when conservatives or nonpartisans report information contrary to the liberal agenda, the liberal journalists cry foul, accusing those who are off narrative of being biased or discredited.

Some Recommendations

Amid the frustration and confusion, amid the slant and spin, you can find reporters providing facts and truth. Here I have gathered a few examples of important work being done by alternative media, non-mainstream reporters, off-narrative journalists, bloggers, partisans, and citizen journalists. I've also included a sampling of national mainstream news reporters who do good, reliable work in today's crazy media environment. For help in compiling this list, I asked numerous colleagues for their recommendations. You may not agree with all of the selections, and it's far from a comprehensive list, but I think it is a decent starting point for people looking for leads on where to seek off-narrative information. The conservative press is breaking stories, but important stories are also being unearthed by some in the liberal press who never stopped watchdogging the government and have resisted partisan interests and bullying by colleagues.

Organizations and Publications

The Epoch Times

This newspaper has conducted impressive in-depth investigations on off-narrative topics. It covers a wide range of national and international news that is overlooked elsewhere. You'll find the news presented in a way that is factual and fair. For example, the publication dug deep into Special Counsel Robert Mueller's Trump-Russia probe without getting bogged down in political spin. It produced the most in-depth graphic representation that I saw of abuse involving the Foreign Intelligence Surveillance Court. It also resisted prevailing narratives about coronavirus to report accurately on theories possibly connecting the 2020 coronavirus outbreak to a Chinese research lab.

RealClearPolitics

This website compiles articles and editorials from all viewpoints left, right, and in between. It also conducts its own original off-narrative, fearless investigations under the title "RealClearInvestigations." For example, it exposed the fact that the Justice Department has a history of declining to prosecute its own officials, even when the inspector general recommends bringing charges.

Just the News

This website tackles news stories of the day the old-fashioned way without shoving a reporter's views down your throat. It also reports fairly on topics that other news organizations avoid or shun.

The Hill

The news side of this popular website tends to tilt left, but the opinion side features a wide sampling of views and analysis for people who want to learn what all sides are buzzing about.

Vice News

Though generally approaching news from a distinctly left-wing point of view, Vice has produced important and courageous off-narrative programming and documentaries. HBO canceled its seven-year-long relationship with Vice News in 2019, but as of this writing, Vice News continues on as a website and has produced a documentary video series. One episode, "A Middle East Divided," examined "escalating tensions in the world's most volatile region."

Project Veritas

James O'Keefe's creation is often dismissed as conservative and his undercover tactics criticized. However, he has successfully exposed important information on numerous untouchable subjects such as media bias inside CNN and ABC and Planned Parenthood's sale of aborted fetus parts.

CNBC, Fox Business News, Bloomberg

When I want to see what else is going on in the world besides the same few stories repeated on cable news, I often find myself tuning in to business news channels and websites. Here one can sometimes get a broader, fairer overview of world and national news.

WikiLeaks

Though its future is unclear due to efforts to prosecute and punish its founder, Julian Assange, WikiLeaks has proven to be one of the most reliable sources when it comes to original documents such as intelligence community material and, of course, embarrassing and revealing internal emails among shakers and movers within the Democratic Party during the 2016 presidential campaign.

The Intercept

Glenn Greenwald, a journalist and constitutional lawyer, is one of three co—founding editors of this off-narrative, left-leaning website. Greenwald broke the story in 2013 of National Security Agency (NSA) whistleblower Edward Snowden and the controversial hidden surveillance practices of the US intelligence community. Though The Intercept typically reports from the liberal side, it often takes on fellow journalists, liberals, and establishment types when they need oversight. One investigation examined the media scandal embedded within the story of improper FBI spying on US citizens.

Judicial Watch

This conservative-leaning public interest group has had more success suing Democratic and Republican administrations for documents and exposing a wide range of government misdeeds than any other single group I can think of.

The *Wall Street Journal*

A colleague with a *New York Times* background told me he considers the *Wall Street Journal* the best news outlet. "They have the strictest standards when it comes to keeping opinion out of their news accounts," he said, citing its outstanding investigations into the workings of Amazon and Google as examples. I often find the *Journal* off narrative when compared to other news outlets.

OpenSecrets.org

The Center for Responsive Politics, which runs the website Open Secrets.org, amasses an incredible amount of data on political spending and trends regarding virtually every political donor, whether individual or corporate, and all federal candidates, whether Democrat, Republican, or other.

People

Pete Williams, NBC News justice correspondent

A longtime colleague describes the strength of Williams's reporting: "The single best example of nonpartisan, no-idea-where-he-stands (even though he once worked for a Republican administration), yet he *never* strays beyond the story and the facts known at the time. He's broken some big stories. But just as significantly, he's pushed back when the momentum of the consensus reporting says one thing. Williams has been known to say 'I can't confirm' or, even braver, 'Everyone else is wrong.' He gained a lot of attention covering the Boston Marathon bombing in 2013 more accurately than others, showing what *The Atlantic* referred to as 'restraint in not jumping too far into conclusions.'"

David Martin, CBS News national security correspondent

Martin has covered the Pentagon and the State Department since 1993. What's true of Pete Williams can also be said of Martin, with whom I worked at the CBS Washington bureau for two decades. He proved to be a steady hand, a fair arbiter, and a generous colleague. I don't know where he stands politically. Well connected, he's respected enough that he gets inside information without having to kowtow to any member of the establishment, whoever it might be.

Kerry Sanders, NBC News Miami-based correspondent

Sanders is an apolitical figure who has covered everything from politics and turmoil in Cuba to wars and domestic features during his more than twenty years at NBC News. I worked with Sanders when we were both in local news in Tampa, Florida. He is driven by curiosity, open-mindedness, fairness, and an eagerness to deliver interesting, off-narrative information in an unbiased fashion.

Mark Knoller, former CBS News Radio White House correspondent
Knoller was always one of my favorite Twitter follows. He tends to stick to the facts and is a walking encyclopedia when it comes to memory about the White House. Prior to publication of this book, it was announced he had been laid off at CBS.

Diane Sawyer, ABC News
"Diane Sawyer has done some great documentaries," says a news executive who highlighted Sawyer as a standout among the many high-profile news personalities he's worked with for more than four decades at the television news networks.

Greta Van Susteren
Whatever Van Susteren may be up to in terms of reporting, I like to watch. She provides smart legal and news analysis and works hard to listen to various viewpoints without grinding axes.

Kim Strassel, Mollie Hemingway, Sara Carter, Gregg Jarrett
These conservative-leaning reporters and analysts frequently break important news that cuts against the grain. Jarrett, who is an attorney, conducted meticulous dissections of the Trump-Russia probe and FBI misbehavior.

Peter Schweizer, investigative journalist and author
Approaching political investigations from the right, Schweizer has conducted important, in-depth investigations into the Clinton Foundation and other "untouchable" subjects that the mainstream media couldn't bring themselves to dive into.

T. Christian Miller, ProPublica; Dave Levinthal, Business Insider
It can be argued that ProPublica and Levinthal's alma mater, the Center for Public Integrity, have some issues with bias, but Miller and

Levinthal, now at Business Insider, are among a number of reporters at those organizations who do strong work. Formerly of the *Los Angeles Times*, Miller coauthored the rape investigation series that won a Pulitzer Prize and was turned into the widely watched Netflix series *Unbelievable*. The Center for Public Integrity's recent political coverage has earned two Edward R. Murrow Awards and two National Headliner Awards, among many other honors.

Andy Pasztor, the *Wall Street Journal*
Pasztor has covered Boeing and aviation for decades. He has won the admiration and respect of many colleagues. "His work on the 737 Super Max has been outstanding. Very authoritative," said one national journalist of Pasztor.

Jeff Gerth
A Pulitzer Prize–winning reporter who has worked for the *New York Times* and ProPublica, Gerth has spent decades defying narratives and beating back intimidation by powerful interests. His wide-ranging investigations have included the transfer of American satellite-launch technology to China and the many Clinton scandals.

Don Van Natta, Jr., ESPN
Formerly of the *New York Times*, Van Natta is described by one colleague as "thorough, hard-working, and fair. . . . [He has] done great investigations on the NFL, Roger Goodell, Venus Williams and the US Tennis Association, and more."

Gretchen Morgenson, the *Wall Street Journal*
A Pulitzer Prize–winning financial reporter, Morgenson earned the respect of fellow journalists for her coverage of Wall Street at the *New York Times*.

James Grimaldi, the *Wall Street Journal*

Formerly of the *Washington Post*, Grimaldi is a Pulitzer Prize–winning journalist who has provided insightful and important coverage of corporate and government misdeeds, among other topics.

Howie Kurtz, Fox News media critic

Kurtz bends over backward to represent varied views and listen to different sides of the story. His Sunday program, *MediaBuzz*, tackles the day's news from a media viewpoint in a more comprehensive and fairer way than many news programs do.

James Rosen

Now working with me at Sinclair Broadcast Group, this former Fox News reporter examines national and political issues with a clear eye. He considers opposing viewpoints fairly rather than forcing any particular ideology down the viewer's throat.

John Solomon

Solomon became a national target of the mainstream press after he broke numerous stories about Ukraine's alleged interference in the 2016 election—a story many powerful people, including some in the media, worked hard to deny and bury. Solomon has sources in high places among both Democrats and Republicans, and he isn't afraid to go off narrative.

Lara Logan

Lara Logan, formerly of CBS News, has proved to be an important voice on media bias. She draws constant flak for her investigations that cut across the mainstream grain and defy The Narrative.

Conclusion

Once again, as I put the finishing notes to a book, it strikes me that the subject I began writing about two years ago happens to be front and center in the national conversation today as the book goes to publication. It's not that I can tell the future. I'm simply a bit of the canary in the coal mine, as I have mentioned. Or maybe, more accurately, I'm a test case. The type of reporting I do makes me an obvious target for those deploying the latest tricks and tactics in their mission to controversialize inconvenient facts, or even hide them entirely from public view.

I remember first identifying these strategies and some of the key players about twenty years ago as I investigated a diverse set of topics for CBS News. Now it's difficult for me to remember a time when I was surprised by the character assassinations, the assaults on factual reporting, the dishonesty, and the attempts to silence entire lines of thought. Back then, of course, I had not figured out what was driving all this behavior. I only knew that certain organized responses to particular news stories made no logical sense.

I closely observed, and even began to study, as the tactics pioneered by pharmaceutical interests and crisis management PR firms were adopted and perfected by other corporate players, political groups, and nonprofits. In some instances, their efforts have proved quite successful. They can hijack the public discussion to such a degree that their slanted views become widely accepted as "truth" that is not open to debate.

But I've also seen an increasingly informed subset of the public who have made it their business to dig a little deeper, follow the money, and think for themselves. If you're reading this book, you're likely among them. You've figured out what's going on. But what about the rest? Are they simply tuning out? As we've seen, that tends to benefit the propagandists.

I first viewed this book as the third in a trilogy I've written addressing sea changes in the news and media. But I now see it's about something much larger than that. What I once viewed as a fundamental transformation of the news industry is really part of a political, societal, and cultural transformation. I no longer see a way to separate these items into distinct animals. As much growth as I've seen in aggressive efforts to censor, control, and manipulate information, and to use the news to forcibly shape public opinion, I've heard the same complaints coming from members of Congress and their staff; from workers at federal agencies; and from corporate employees. They say independent watchdogging is discouraged. If facts are contrary to a powerful interest, they are to be buried. Views that don't line up with the preferred narrative are filtered out, and messengers of independent information are smeared and destroyed. Congressional hearings are often just for show with little follow-up. More often than not, political leaders ensure there are no congressional hearings at all on certain controversies. Those who raise their hands about wrongdoing inside their organizations, even when the wrongdoing puts public lives at risk, are often treated as if they are traitors to be silenced, punished, and destroyed. On the other hand, any message that's friendly to the cause—usually determined on the basis of who is paying money to whom somewhere—is delivered through every possible medium. As I've traveled to Europe, Asia, and Russia, I've observed similar trends. It's gone global. Journalism and the media aren't what they used to be. They're tools that powerful players use to dictate what others may know and think.

Despite being force-fed information today, we cognitively know information is constantly changing and ever expanding. Different authorities have differing interpretations of the same material. The public has a variety of viewpoints on many topics, and understands the special nature of the American spirit where we are free to say, think, and conclude what we wish, free from government or outside censorship (except concerning that which is deemed illegal). The confounding factor is: the majority who understand these things typically believe in following rules and laws, and eschew violence. So, how can they be any match for a vocal, well-funded minority, supported by key media entities, who blatantly circumvent rules and laws, and may successfully impose their ways using violence—often without punishment? Under that scenario, the same side always wins.

What do the information dictators ultimately hope to achieve? A managed information landscape where they don't really even have to step in and censor; it becomes reflexive. Where we self-censor our thoughts and deeds. Where we know what's permitted and what is not. We fall in line with what they want us to say and think—or we pretend to because resistance is pointless. There are already many people who have acclimated to, or even embraced, this new reality. It sneaked up on them, and once they recognized it, they were already in too deep to reverse course. *Limit what knowledge I'm allowed to find on the Internet? Well, what can I do about that? "Curate" my information? Fine. I'm all in.*

These are dangerous times in terms of what we may become. As clearly as Orwell laid out his nightmarish version of the future—we seem to be arriving there all the same. In an irony he surely would appreciate, his books may one day be disappeared, as if they never existed.

So, as more people come to recognize what's happening, is it already too late to turn the ship?

All of this is not to say that "the news" should be slanted in a different direction. And it's not to suggest we can go back to the way things were—or that it was a perfect model for what should be. But I'd like to believe there is a version of our future in which information is accessible in its many forms, with the recognition that often what's right or wrong, and what's considered factually correct, is no more than a matter of opinion or a snapshot in time. Where we are invited to use our own brains to think what we like, form our own conclusions, change our minds, feel out our positions, argue, and debate—free from the grip of political and corporate interests or social activists who increasingly seek to limit what we may know and say. The quest for knowledge is ongoing and never final.

I choose to believe there is a viable path to such a place because the alternative is too chilling. In an alternate future, people will be told this book was never written.

Appendix:
Major Media Mistakes
in the Era of Trump

AUGUST 2016–JUNE 2020

(The ongoing list can be found at https://sharylattkisson.com under Special Investigations > "Media Mistakes in the Trump Era: The Definitive List")

1. AUGUST 2016–NOVEMBER 2016: Various news outlets publish modeling photos of Trump's wife, Melania, implying that she violated her visa status as an immigrant. But the media got the date wrong.

2. OCTOBER 1, 2016: The *New York Times* and other media imply Trump did not pay income taxes for eighteen years. But tax returns later leaked to MSNBC show Trump actually paid a higher rate than Democrats Bernie Sanders and President Barack Obama.

3. OCTOBER 18, 2016: In a *Washington Post* piece not labeled opinion or analysis, Stuart Rothenberg incorrectly reports that Trump's path to an electoral college victory is "nonexistent."

4. NOVEMBER 4, 2016: *USA Today* "misstates" Melania Trump's arrival date from Slovenia amid a flurry of reporting questioning her immigration status from the mid-1990s.

5. NOVEMBER 9, 2016: Early on election night, the *Detroit Free Press* calls the state of Michigan for Hillary Clinton. (Trump actually won Michigan.)

6. JANUARY 20, 2017: CNN claims Nancy Sinatra was "not happy" about her father's song being used at Trump's inauguration. Sinatra responds, "That's not true. I never said that. Why do you lie, CNN? Actually I'm wishing him the best."

7. JANUARY 20, 2017: Zeke Miller of *Time* reports that President Trump has removed the bust statue of civil rights leader Martin Luther King, Jr., from the Oval Office. The news goes viral. It is false.

8. JANUARY 26, 2017: Josh Rogin of the *Washington Post* reports that the State Department's "entire senior administrative team" has resigned in protest against Trump. A number of media outlets, ranging politically from left to right, state that claim is misleading or wrong.

9. JANUARY 28, 2017: CNBC's John Harwood reports the Justice Department "had no input" into Trump's immigration executive order. Harwood later amends his report to reflect the fact that Justice Department lawyers reviewed Trump's order.

10. JANUARY 31, 2017: CNN's Jeff Zeleny reports the White House set up Twitter accounts for two judges to try to keep their selection for the Supreme Court by Trump secret. Zeleny later corrects his report to state that the allegation was untrue.

11. FEBRUARY 2, 2017: TMZ reports Trump has changed the name Black History Month to African American History Month, implying the change is racist. In fact, Presidents Obama, George W. Bush, and Bill Clinton all previously called Black History Month "African American History Month."

12. FEBRUARY 2, 2017: AP and others report Trump threatened the president of Mexico with invasion to get rid of "bad hombres." The White House says it wasn't true, and the *Washington Post* removes the AP info that "could not be independently confirmed."

13. FEBRUARY 4, 2017: Josh Rogin of the *Washington Post* reports on "Inside the White House—Cabinet Battle over Trump's Immigration Order." The article is repeatedly "updated" to note that one of the reported meetings did not actually occur, a conference call did not happen as

described, and actions attributed to Trump were actually carried out by his chief of staff.

14. FEBRUARY 14, 2017: The *New York Times'* Michael S. Schmidt, Mark Mazzetti, and Matt Apuzzo report on supposed contacts between Trump campaign staff and "senior Russian intelligence officials." FBI director James Comey later testifies, "In the main, [the article] was not true."

15. FEBRUARY 22, 2017: ProPublica's Raymond Bonner reports CIA official Gina Haspel, Trump's later pick for CIA director, was in charge of a secret CIA prison where Islamic extremist terrorist Abu Zubaydah was waterboarded eighty-three times in one month and that she mocked the prisoner's suffering. More than a year later, ProPublica retracts the claim, stating that "Neither of these assertions is correct. . . . Haspel did not take charge of the base until after the interrogation of Zubaydah ended."

16. APRIL 5, 2017: An article by the *New York Times'* graphic editors Karen Yourish and Troy Griggs refers to Trump's daughter Ivanka as Trump's wife.

17. MAY 10, 2017: Numerous outlets, including Politico, the *New York Times*, the *Washington Post*, CNN, AP, Reuters, and the *Wall Street Journal*, report the same leaked information: that Trump fired FBI director Comey shortly after Comey requested additional resources to investigate Russian interference in the election. The Justice Department, Deputy Attorney General Rod Rosenstein, and Acting FBI Director Andrew McCabe say the media reports were untrue, and McCabe adds that the FBI's Russia investigation was "adequately resourced."

18. MAY 27, 2017: The BBC's James Landale, *The Guardian*, and others report that Trump didn't bother to listen to the translation during a speech in Italian by Italy's prime minister. After the reports circulated, the White House states that, as always, Trump was indeed wearing a translation earpiece in his right ear.

19. JUNE 4, 2017: NBC News tweets that Russian president Vladimir Putin told TV host Megyn Kelly that he has compromising information about Trump. Actually, Putin said the opposite: that he does not have compromising information on Trump.

20. JUNE 6, 2017: CNN's Gloria Borger, Eric Lichtblau, Jake Tapper, and Brian Rokus and ABC's Justin Fishel and Jonathan Karl report that FBI director Comey was going to refute Donald Trump's claim in congressional testimony that Comey told Trump three times he was not under investigation. Instead, Comey confirmed Trump's claim.

21. JUNE 7, 2017: In a fact-check story, AP erroneously reports that Trump misread the potential cost to a family with insurance under the Affordable Care Act who wanted care from their existing doctor.

22. JUNE 8, 2017: The *New York Times*' Jonathan Weisman reports FBI director Comey testified behind closed doors that Attorney General Jeff Sessions, a Trump appointee, had told him not to call the Russia probe "an investigation" but "a matter." Weisman was mistaken. Actually, it was Attorney General Loretta Lynch, an Obama appointee, not Sessions, who told Comey to refer to the Hillary Clinton classified email probe (not the Russia probe) as "a matter" instead of "an investigation."

23. JUNE 22, 2017: CNN's Thomas Frank reports that Congress is investigating a "Russian investment fund with ties to Trump officials." The report is later retracted. Frank and two other CNN employees resign in the fallout.

24. JULY 6, 2017: *Newsweek*'s Chris Riotta and others report Poland's first lady refused to shake Trump's hand. *Newsweek*'s later "update" reflects the fact that the first lady shook Trump's hand after all, as clearly seen on the full video.

25. JULY 6, 2017: The *New York Times*' Maggie Haberman, CNN, and numerous outlets have long reported, as if it were fact, Hillary Clinton's claim that a total of seventeen American intelligence agencies concluded that Russia had orchestrated election-year attacks to help get Trump elected. Only three or four agencies, not seventeen, had officially done so.

26. AUGUST 31, 2017: NBC News' Ken Dilanian and Carol E. Lee report that a Trump official's notes about a meeting with a Russian lawyer included the word "donation," as if there had been discussions about suspicious campaign contributions. NBC later corrects the report to reflect that

the word "donation" didn't appear, but it still claims the word "donor" did. Later, Politico reports the word "donor" wasn't in the notes, either.

27. SEPTEMBER 5, 2017: CNN's Chris Cillizza and reporters at other news outlets declare that Trump "lied" when he stated that Trump Tower had been wiretapped, although there's no way any reporter could independently know the truth of the matter, only what intel officials claimed. It later turns out there were numerous wiretaps involving Trump Tower, including during a meeting of Trump officials with a foreign dignitary. At least two Trump associates who had offices in or frequented Trump Tower were also reportedly wiretapped.

28. SEPTEMBER 7, 2017: The *New York Times*' Maggie Haberman reports that Democrat House speaker Nancy Pelosi called President Trump about an immigration issue. Actually, Trump made the call to Pelosi.

29. NOVEMBER 6, 2017: CNN's Daniel Shane presents edited excerpts from a Trump event to make it seem as though Trump didn't realize that Japan builds cars in the United States. However, Trump's full statement makes clear that he does.

30. NOVEMBER 6, 2017: CNN edits a video to make it appear as though Trump impatiently dumped a box of fish food into the water while feeding fish at the Imperial Palace in Tokyo, Japan. The New York *Daily News*, *The Guardian*, and others publish stories implying Trump is gauche and impetuous. The full video shows Trump simply followed the lead of Japan's prime minister.

31. NOVEMBER 29, 2017: *Newsweek*'s Chris Riotta claims Ivanka Trump "plagiarized" one of her own speeches. In fact, plagiarizing one's own work is impossible since plagiarism occurs when a writer steals someone else's work and passes it off as his own.

32. DECEMBER 2, 2017: ABC News' Brian Ross reports former national security advisor Lieutenant General Michael Flynn, a Trump appointee, is going to testify that candidate Trump directed him to contact "the Russians." Even though such contact would not be in and of itself a violation of the law, the news was treated as an explosive indictment of Trump in the Russia collusion narrative, and the stock market fell on the news.

ABC later corrects the report to reflect that Trump had already been elected when he reportedly asked Flynn to contact the Russians about working together to fight ISIS and other issues. Ross is suspended.

33. DECEMBER 4, 2017: The *New York Times'* Michael S. Schmidt and Sharon LaFraniere, as well as other journalists, report that Deputy National Security Advisor K. T. McFarland, a Trump appointee, supposedly contradicted herself or lied about another official's contacts with Russians. CNN, MSNBC, CBS News, the New York *Daily News*, and The Daily Beast pick up the story about McFarland's "lies." The story is later repeatedly and heavily amended.

34. DECEMBER 4, 2017: ABC News' Trish Turner and Jack Date report that former Trump campaign chairman Paul Manafort recently worked with a Russia intelligence-connected "official." But the Russian wasn't an "official."

35. DECEMBER 5, 2017: Bloomberg's Steven Arons and the *Wall Street Journal's* Jenny Strasburg report the "blockbuster" that Special Counsel Robert Mueller has subpoenaed Trump's bank records. It isn't true.

36. DECEMBER 8, 2017: CNN's Manu Raju and Jeremy Herb report that Donald Trump, Jr., conspired with WikiLeaks in advance of the publication of damaging Democratic Party and Clinton campaign emails. Many other publications follow suit. They have the date wrong: WikiLeaks and Trump Junior were in contact after the emails were published, not before.

37. JANUARY 3, 2018: Talking Points Memo's Sam Thielman reports that a Russian social media company provided documents to the Senate about communications with a Trump official. The story is later corrected to say the reporter actually had no idea how the Senate had received the documents and had no evidence to suggest the Russian company was cooperating with the probe.

38. JANUARY 12, 2018: Mediaite's Lawrence Bonk, CNN's Sophie Tatum, *The Guardian*, BBC, *U.S. News & World Report*, Reuters, and BuzzFeed's Adolfo Flores report a "bombshell": that President Trump has backed down from his famous demand for a wall along the entire southern US border. However, Trump said the very same thing in February 2016 on

MSNBC; on December 2, 2015, in the *National Journal*; in October 2015 during the CNBC Republican primary debate; and on August 20, 2015, on Fox Business' *Mornings with Maria*.

39. JANUARY 15, 2018: AP's Laurie Kellman and Jonathan Drew report that a new survey shows trust in the media has fallen during the Trump presidency. But the survey that AP cited was actually over a year old and was conducted while Barack Obama was president.

40. JANUARY 31, 2018: Media reports in December 2017 claimed the Trump administration had banned officials at the Centers for Disease Control and Prevention (CDC) from using seven words. In response, doctors posted photos of themselves with tape over their mouths. It turned out the documents showed there was "not a ban or prohibition on words but rather suggestions on how to improve the chances of getting funding."

41. FEBRUARY 2, 2018: AP's Eric Tucker, Mary Clare Jalonick, and Chad Day report that ex–British spy Christopher Steele's opposition research against Trump was initially funded by a conservative publication: the *Washington Free Beacon*. AP corrects its story because Steele came on the project only after Democrats began funding it.

42. MARCH 8, 2018: The *New York Times*' Jan Rosen reports on a hypothetical family whose tax bill would rise nearly $4,000 under Trump's tax plan. It turns out that the couple's taxes would actually go down by $43, not up by $4,000.

43. MARCH 13, 2018: The *New York Times*' Adam Goldman, NBC's Noreen O'Donnell, and AP's Deb Riechmann report that Trump's pick for CIA director, Gina Haspel, waterboarded a particular Islamic extremist terrorist dozens of times at a secret prison and that she mocked his suffering. In fact, Haspel wasn't assigned to the prison until after the detainee left. (ProPublica originally reported the incorrect story in February 2017.)

44. MARCH 15, 2018: AP's Michael Biesecker, Jake Pearson, and Jeff Horwitz report that a Trump advisory board official was a Miss America contestant and killed a black rhino. She was actually a Mrs. America contestant and shot a nonlethal tranquilizer dart at a white rhino.

45. APRIL 1, 2018: AP's Nicholas Riccardi reports the Trump administration has ended a program to admit foreign entrepreneurs. It isn't true.

46. APRIL 30, 2018: AP reports the NRA has banned guns during speeches by Donald Trump and Mike Pence at the NRA's annual meeting. AP later corrects the information because the ban was put in place by the Secret Service.

47. MAY 3, 2018: NBC's Tom Winter reports the government wiretapped Trump's personal attorney Michael Cohen. NBC later corrects the story after three senior US officials say there was no wiretap.

48. MAY 7, 2018: CNBC's Kevin Breuninger reports that Trump's personal lawyer Michael Cohen paid $1 million in fines related to unauthorized cars in his taxi business, was barred from managing taxi medallions, had transferred $60 million offshore to avoid paying debts, and is awaiting trial on charges of failing to pay millions of dollars in taxes. A later correction states that none of that is true.

49. MAY 16, 2018: The *New York Times*' Julie Hirschfeld Davis, AP, CNN's Oliver Darcy, and others excerpt a Trump comment as if he had referred to immigrants or illegal immigrants in general as "animals." Most outlets later correct their reports to note Trump specifically referred to members of the murderous criminal gang MS-13.

50. MAY 28, 2018: *The New York Times Magazine*'s editor in chief, Jake Silverstein, and CNN's Hadas Gold share a story with photos of immigrant children in cages as if they were new photos taken during the Trump administration. The article and photos were in fact from 2014 during the Obama administration.

51. MAY 29, 2018: The *New York Times*' Julie Hirschfeld Davis reports the estimated size of a Trump rally to be 1,000 people. There were actually 5,500 people or more in attendance.

52. JUNE 1, 2018: In a story about Trump tariffs, AP reports the dollar value of Virginia's farm and forestry exports to Canada and Mexico was $800. It was actually $800 million.

53. JUNE 21, 2018: *Time* magazine and others use a photo of a crying Honduran child to illustrate a supposed Trump administration policy separating illegal immigrant parents and children. The child's father later reports that agents never separated her from her mother; the mother had taken her to the United States without his knowledge and separated herself from her other children, whom she had left behind.

54. JUNE 22, 2018: An MSNBC personality mistakenly states Trump has "banned" the Red Cross from visiting children separated from illegal immigrant parents.

55. JUNE 28, 2018: After a newsroom shooting, a newspaper reporter tweets that the shooter "dropped his [Make America Great Again] hat on newsroom floor before opening fire." The story was false.

56. JULY 10, 2018: NBC reporter Leigh Ann Caldwell incorrectly reports that outgoing Supreme Court justice Anthony Kennedy retired only after months of negotiations with Trump that concluded with Trump agreeing to replace Kennedy with Brett Kavanaugh.

57. JULY 16, 2018: After Trump discusses Finland in regard to a NATO meeting, a *Washington Post* reporter implies that Trump doesn't know that Finland is not a NATO country. In fact, Trump met with the Finnish president at the NATO summit. Further, Finland is a NATO partner.

58. SEPTEMBER 14, 2018: The *New York Times* issues a major correction to an original "unfair" article about US ambassador to the United Nations Nikki Haley.

59. SEPTEMBER 18, 2018: The *New York Times* reports that a man named Mark Judge testified he remembered an incident more than thirty years before in which Supreme Court nominee Brett Kavanaugh was accused of assault. Judge actually said the opposite: that he does not remember such an incident and the allegations are "absolutely nuts."

60. SEPTEMBER 23, 2018: Multiple news outlets report that Deputy Attorney General Rod Rosenstein has resigned or been fired. Neither turns out to be true. Axios and others eventually "update" and "clarify" their erroneous reports.

61. OCTOBER 14, 2018: NBC News reports that President Trump praised Confederate general Robert E. Lee. Actually, Trump praised the Union general Ulysses S. Grant.

62. NOVEMBER 14, 2018: CNN's Jeff Zeleny reports President Trump has decided to fire a deputy national security advisor upon the first lady's urging. The *Wall Street Journal* reports the adviser was "escorted out" of the White House. Later, it's reported that neither was true. "This did not happen. She is still here at the WH," a senior official tells the press. (The adviser was reassigned to another job.)

63. DECEMBER 24, 2018: It's discovered that nearly everything written by a *Der Spiegel* reporter who has been honored by CNN about a supposedly racist Trump stronghold town had been fabricated—like much of his other work.

64. DECEMBER 26, 2018: NBC reports Trump was the first president since 2002 not to visit the troops at Christmastime. But he (and First Lady Melania) did do so. NBC later adds a note to its story but leaves the false headline in place.

65. JANUARY 2019: The *New York Times*, Vice, and others report on the "lost" immigrant children of the Trump administration. However, AP and other fact-checkers state this is a misleading term: the "lost" children were a matter of the government not being able to track them once they were placed with sponsors who themselves were often in the United States illegally. According to AP, in the last three months of 2017, the Trump administration slightly exceeded the success rate of the Obama administration when it comes to tracking the children.

66. JANUARY 1, 2019: CBS News claimed, in June 2018, that Trump spokesman Sarah Huckabee Sanders would retire by the end of the year. She didn't. The same CBS story quoted sources as saying the departure of White House deputy assistant to the president Raj Shah was also imminent. It wasn't. (Shah continued to serve for seven more months.) The *Washington Post* and others reported last November that Trump was imminently about to fire DHS secretary Kirstjen Nielsen. The *Post* confirmed this with five anonymous sources. The firing was said to be likely to happen the following week. However, she remained at work for five more months.

67. JANUARY 9, 2019: The *New York Times* issues a correction to a report that falsely stated former Trump campaign chairman Paul Manafort once asked for campaign polling data to be given to a Russian oligarch who has ties to Russian president Vladimir Putin. Instead, the *Times* now claims, Manafort actually asked his associate Rick Gates to give polling data to Ukrainian oligarchs—not the Russians.

68. JANUARY 11, 2019: The Fox TV affiliate in Seattle, Washington, airs fake, doctored video of President Trump that alters his face and makes it appear as though he stuck his tongue out and pulled it in while giving an Oval Office address.

69. JANUARY 18, 2019: The BuzzFeed exclusive with anonymous sources implicating Trump in potentially criminal behavior is refuted in a rare rebuke from Special Counsel Mueller's office.

70. JANUARY 22, 2019: The *New York Times* and *Washington Post* are among the publications that issue corrections after falsely reporting that an anti-Trump activist served in the Vietnam War. Additionally, several news employees, including a CNN employee, apologize for mischaracterizing, as the aggressors, Trump-supporting teenagers at a pro-life rally.

71. JANUARY 26, 2019: The United Kingdom's *Telegraph* apologizes for all the facts it got wrong in a January 19 article criticizing the first lady.

72. FEBRUARY 18, 2019: Some media outlets unskeptically further the false narrative that the actor Jussie Smollett, who is black, was attacked by Trump-supporting racists who put a noose around Smollett's neck, shouted racial slurs, told him "It's MAGA [Make America Great Again] country," and poured bleach on him.

73. FEBRUARY 27, 2019: McClatchy and others report that former Trump lawyer Michael Cohen visited Prague to meet with Russians to help collude on Trump's behalf. Cohen later testifies to Congress that he's never been to Prague or the Czech Republic.

74. MARCH 1, 2019: The *Washington Post* deletes a tweet containing false reporting about a January 19 incident regarding a standoff between Trump-supporting pro-life Catholic high school students and a pro-choice Native American activist. The *Post* wrongly stated, without attribution, that the activist had fought in the Vietnam War. The activist also falsely stated that a high school student had blocked him and "wouldn't allow him to retreat."

75. APRIL 2019: The release of Special Counsel Robert Mueller's report on Trump-Russia collusion contradicts multiple reporters and media outlets that falsely reported on the timing. The *Washington Post* said the report would be out in the summer of 2018. Bloomberg said it would be out shortly after the 2018 midterm elections. In February 2019, CNN, The *Washington Post*, and NBC reported it would be out in the last week of February.

76. MAY 29, 2019: The *Wall Street Journal* reports that the Navy hung a tarp to cover the name of USS *John S. McCain* so President Trump wouldn't see it on his visit to Yokosuka, Japan. It is further reported the ship was kept out of Trump's view and sailors wearing hats with the ship's name on it were turned away and/or given the day off so that Trump would not see the McCain name. A military official did send an email directing that USS *McCain* be kept from Trump's view. However, that direction was not followed; Trump and White House aides indicate Trump played no role and was unaware of the direction; military officials state it is untrue that a tarp was placed over the ship's name to block it from Trump's view; in fact a tarp on the ship for maintenance was removed for Trump's visit.

77. JULY 4, 2019: Several news outlets report that President Trump's Fourth of July celebration did not draw crowds or drew "small crowds." However, by any factual assessment, the crowds were, in fact, huge. It turns out *The Guardian* had featured a misleading photo taken prior to the event.

78. JULY 13, 2019: In a story about a lawsuit alleging that candidate Trump forcibly kissed a campaign worker, CNN fails to mention the lawsuit was dismissed. CNN later corrects the story.

79. JULY 21, 2019: Many in the media uncritically report an African American Georgia state legislator's claim that a white man at a grocery store told her to "go back where you came from." Media reports link the supposed hateful comment to President Trump because Trump recently said several Democrats in Congress should "go back and help fix the totally broken and crime infested places from which they came." However, the following day, the legislator acknowledges the man did not say she should "go back to your country" or "go back to where you came from," as she originally claimed. She goes on to admit *she* was the one who told *him* to "go back." The man adds he is not white but a Cuban and a Democrat.

80. JULY 21, 2019: An MSNBC contributor and law professor tweets that Fox is not planning to air upcoming congressional testimony by former special counsel Robert Mueller on the Trump-Russia investigation. When the error is pointed out, the contributor says she was just kidding.

81. JULY 24, 2019: In testimony to Congress, Special Counsel Robert Mueller puts to final rest the widespread reporting originating with Slate in 2016 that claimed a Russian bank server had been illicitly communicating with Trump Tower. When asked about it by a member of Congress, Mueller replied: "My belief at this point is . . . not true."

82. JULY 29, 2019: Vox's Aaron Rupar tweets that Trump suggested he was a "9/11 First Responder." In fact, Trump stated, "I'm not considering myself a first responder."

83. AUGUST 2019: Multiple news outlets, including CNN and MSNBC, falsely report that an illegal immigrant had her nursing baby ripped from her arms. CNN later acknowledges the mother was not lactating and was not nursing.

84. AUGUST 5, 2019: MSNBC's Nicolle Wallace falsely claims President Trump talked about "exterminating Latinos." She apologizes the next day, tweeting "I misspoke about Trump's calling for an extermination of Latinos. My mistake was unintentional and I'm sorry."

85. AUGUST 28, 2019: MSNBC's Lawrence O'Donnell apologizes for and retracts anonymous, unverified claims stating that Trump took out loans with Russian cosigners.

86. AUGUST 28, 2019: Ken Dilanian of NBC News corrects a false report he and others disseminated claiming that starting October 29, "children born to U.S. service members outside of the U.S. will no longer be automatically considered citizens. Parents will have to apply for citizenship for their the [*sic*] children in those situations."

87. SEPTEMBER 7, 2019: CNN and nearly every other major media outlet criticize President Trump for tweeting that Alabama will likely be impacted by Hurricane Dorian, saying that was ridiculous. However, multiple official hurricane advisories had put Alabama into a projected impacted area.

88. SEPTEMBER 10, 2019: Citing anonymous sources, CNN and the *New York Times* report—and other media repeat—claims that the CIA had to remove a top US spy from Russia in 2017 because of concern over President Trump's handling of classified information. The CIA, Secretary of State Mike Pompeo, and the White House strongly refute the story. Other media also contradict CNN and report the decision to remove the spy happened before CNN said it did and for different reasons.

89. SEPTEMBER 16, 2019: The *New York Times* publishes an editor's note about its recent story recounting a newly reported accusation about an incident decades ago involving Supreme Court Justice Brett Kavanaugh. The editor's note discloses for the first time that the *Times* never spoke to the alleged victim and that the alleged victim had told friends she had no recollection of any such event.

90. SEPTEMBER 25, 2019: The *Washington Post*, quoting anonymous sources, reports President Trump's director of national intelligence, Joseph Maguire, threatened to quit over an alleged issue. However, Maguire issues the statement, "At no time have I considered resigning my position since assuming this role."

91. SEPTEMBER 25, 2019: The Daily Beast and other media outlets report President Trump asked the president of Ukraine to investigate former vice president Joe Biden's son Hunter eight times in one phone call. However, the released transcript notes reveal Trump mentioned Biden's son (and not by name) one time.

92. SEPTEMBER 29, 2019: Scott Pelley of CBS News' *60 Minutes* reports, "The government whistleblower who set off the impeachment inquiry of President Trump is under federal protection because they fear for their safety." But the attorney for the unnamed "whistleblower," Mark Zaid, then tweets a statement reading "NEWS ALERT: 60 Minutes completely misinterpreted contents of our letter."

93. SEPTEMBER 30, 2019: When a twelve-year-old black girl claims white boys at school held her down, cut off her hair, and called her "nappy" and "ugly," the story makes national news. Multiple news outlets improperly report some details as if they are established as true, without proper attribution. For example, NBC writes, "The attack happened Monday" and "The second boy grabbed her arms, while the third cut off some of her dreadlocks." A local NBC affiliate writes, "she was at recess and about to go down a slide when one of the boys grabbed her and put a hand over her mouth. Another boy grabbed her arms. A third boy cut off some of her hair." CBS writes, "The incident took place" (as if an incident had been factually established rather than being an allegation). Many news reports also connect the attack to President Trump's vice president, Mike Pence, by stating that the "attack" happened at "a Christian school in Virginia where Vice President Mike Pence's wife works." However, it turns out there was no attack or "incident." Three days after the initial reports, the child's family reports the whole story was made up and apologizes.

94. OCTOBER 13, 2019: ABC airs video purportedly showing a "slaughter" and "horrific report of atrocities" against Kurds by Turkey after President Trump withdrew US troops. But the video isn't of combat or even in Turkey; it's a file tape of a training video in the United States.

95. OCTOBER 16, 2019: Many major news outlets, including Yahoo!, *USA Today*, Roll Call, NBC, ABC, and Fox, quote President Trump as saying Turkey's invasion of Syria "is not our problem." In a subsequent correction, NBC and others admit it was a misquote: Trump actually said "it's not our border."

96. OCTOBER 27, 2019: Multiple media sources state that President Trump was golfing during the US raid in Syria that captured the head of the

Islamic terrorist group ISIS, Abu Bakr al-Baghdadi; and that a White House situation room photo had been "staged." It turns out Trump had finished golfing and was at the White House during the operation.

97. NOVEMBER 16, 2019: There's rampant speculation after a contributor to The Hill claims President Trump visited Walter Reed National Medical Center due to chest discomfort. A White House statement from Trump's physician states that was not the case: "Despite some of the speculation, the President has not had any chest pain, nor was he evaluated or treated for any urgent or acute issues. Specifically, he did not undergo any specialized cardiac or neurologic evaluations."

98. NOVEMBER 19, 2019: The United Kingdom's *Daily Mail* posts a sensational headline during the impeachment hearings against President Trump claiming Ambassador to NATO Kurt Volker had "walked back" his testimony in a way that was detrimental to Trump. But when Volker was asked at the hearing if the *Daily Mail* headline was correct, he stated it was not.

99. NOVEMBER 19, 2019: Agence France Press publishes sensational "breaking news" that more than 100,000 children are being held in migration-related detention in the United States under President Trump. It turns out that was the number in 2015 under President Obama.

100. NOVEMBER 24, 2019: *Newsweek*'s Jessica Kwong reports on President Trump's tipping at a restaurant, implying he'd been cheap. *Newsweek* later "updates" the story to remove the headline reference to a "thin stack of cash" and include that the stack of cash given in tips was hundred-dollar bills, above what Trump had already tipped the servers.

101. NOVEMBER 28, 2019: Kwong of *Newsweek* falsely reports President Trump is spending Thanksgiving golfing in Florida at his Mar-a-Lago Club. He is actually in Afghanistan serving dinner to US troops.

102. DECEMBER 3, 2019: Congressman Devin Nunes, a Republican from California, files a $4.35 million defamation lawsuit against CNN for claiming he flew to Vienna, Austria, in December 2018 to meet with a former Ukrainian prosecutor in to dig up dirt on Joe Biden and his son. Nunes says he was actually in Benghazi, Libya, and Malta for meetings;

shows dated photos; and says he never met with the prosecutor in Vienna or anywhere else.

103. DECEMBER 9, 2019: Newly public FBI documents prove countless media sources were wrong when they reported the Democrat-funded "dossier" submitted by the FBI to get a wiretap to spy on Trump associate Carter Page was only a "small part" of the wiretap application and that there was evidence that Trump was a "Putin stooge" and coordinating with Russian president Vladimir Putin or Russia.

104. DECEMBER 16, 2019: The news media widely misreport that the report by Department of Justice inspector general Michael Horowitz found "no political bias" in the Russia probe.

105. DECEMBER 25, 2019: The *Wall Street Journal* reports on Trump administration tariff negotiations with China. The Trump administration issues unequivocal denials of the information reported.

106. DECEMBER 27, 2019: The *New York Times* corrects a report it published to demonstrate that Trump voters no longer support him. It turns out that its featured example was a man who had not voted for Trump in the first place.

107. JANUARY 7, 2020: MSNBC wrongly reports up to thirty US deaths after an Iranian rocket attack. In fact, no Americans were killed.

108. JANUARY 16, 2020: MSNBC's John Brennan, former CIA director, reports Trump personally wrote a note requesting that Ukraine's president announce an investigation into possible corruption related to former vice president Joe Biden and his son Hunter. Brennan later says he was mistaken.

109. FEBRUARY 21, 2020: The *New York Times* and multiple other news outlets report on a secret briefing to Congress in which lawmakers were supposedly told Russia was interfering to try to get Trump reelected in 2020. The report is later followed up by stories indicating the warnings may have been "overstated." In fact, officials told CNN, the United States "does not have evidence that Russia's interference this cycle is aimed at reelecting Trump."

110. FEBRUARY 26, 2020: Amid the coronavirus outbreak, numerous media outlets imply or state that President Trump slashed, cut, or gutted the budget for the Centers for Disease Control. In fact, the CDC budget has increased each year.

111. FEBRUARY 28, 2020: Numerous media outlets falsely report President Trump called the coronavirus a "hoax." In fact, the president called the Democrats' politicization of the outbreak a hoax.

112. MARCH 1, 2020: Congressman Nunes announces plans to sue the *Washington Post* over what he says was a false report claiming he went to the White House and talked to President Trump about a congressional briefing by Acting Director of National Intelligence Joseph Maguire about the prospects of Russian interference in the 2020 campaign. Nunes says he never talked to the president about Maguire and did not go to the White House when the *Post* claimed he did.

113. MARCH 5, 2020: The *Washington Post* editorial team says many of the United States' "hundreds of millions" of voters support Joe Biden. But there are not that many voters in the United States. (In 2018, there were 153 million people registered to vote. Tens of millions of them do not vote.)

114. MARCH 15, 2020: An anonymously sourced news report alleges President Trump attempted to bribe a German coronavirus vaccine maker and wants to hoard the vaccine so only Americans will have it. Reuters reports the German Health Ministry confirmed the report. However, the German Health Ministry disputes the Reuters characterization and the Trump administration denies the Reuters report altogether.

115. MARCH 18, 2020: The *New York Times* and Jeremy Peters publish an article including multiple false claims about Sharyl Attkisson and Rob Schneider, claiming they and others have "minimized" coronavirus risks and "insisted" it it is overplayed. In fact, Peters altered an Attkisson quote and made at least nine false claims about her work. Peters also manipulated a Schneider quote and quoted him out of context in order to make it appear as though he had violated recommendations not to eat at restaurants, when he had not. The *Times* issued multiple corrections about its misreporting on Attkisson.

116. MARCH 19, 2020: Jennifer Rubin of the *Washington Post* blames Senate majority leader Mitch McConnell, a Republican, for delaying a coronavirus-related stimulus package vote. The *Post* later issues a correction.

117. MARCH 27, 2020: The *New York Times* issues a correction after falsely reporting the United States was short at least 800,000 ventilators in the coronavirus crisis because a million would be needed and there were only 200,000 on hand. In fact, a study actually projected that a million people might need a ventilator over the course of the pandemic, not at one time.

118. MARCH 28, 2020: The *New York Times* corrects its timeline about the slow implementation of coronavirus testing in the United States. The date of the country's first confirmed case of Covid-19 spreading through travel was almost two weeks later than stated in the original timeline.

119. MARCH 30, 2020: *CBS This Morning* airs a story supposedly showing video of a New York hospital crowded with coronavirus patients. CBS News later issues a correction after viewers recognize it as file tape from Italy.

120. APRIL 6, 2020: CBS News airs social media video of a crying woman who says she was a nurse and quit her job due to not having masks while treating coronavirus patients. CBS later "clarifies" that the nurse acknowledged masks were available while she was in the room treating coronavirus patients.

121. APRIL 6, 2020: Peter Baker, Katie Rogers, David Enrich, and Maggie Haberman of the *New York Times* report, "Trump has seized on [hydroxychloroquine] as a miracle cure." In fact, the day before the article was published, the president repeatedly qualified his support for hydroxychloroquine—as he usually does—and did not call it a miracle cure.

122. APRIL 8, 2020: CBS News again mistakenly uses the file video from a hospital in Italy in a story about coronavirus-overrun Pennsylvania hospitals.

123. APRIL 14, 2020: The US government publicly confirms it is looking into possible links between coronavirus and a research lab in Wuhan,

China. This makes a *Washington Post* report of February 17, 2020, by Paulina Firozi false. She declared the idea of the virus coming from the Wuhan lab had been "debunked."

124. APRIL 15, 2020: A Facebook "science fact-check" incorrectly flags as "false" an *Epoch Times* documentary about the coronavirus's possible link to a Wuhan, China, research lab. It turns out that a reviewer referenced by Facebook's fact-check is a US scientist who has been working at the Wuhan lab.

125. APRIL 22, 2020: Reuters and other news outlets claim President Trump tapped a "former Labradoodle breeder . . . to lead U.S. pandemic task force." They imply the official, Brian Harrison, is unqualified, and blame him for supposedly slowing the US coronavirus response. The stories in multiple outlets appear on the same day. However, Harrison never led the coronavirus task force.

126. APRIL 25, 2020: After MarketWatch and the *Washington Post* report coronavirus stimulus checks may or "will" be delayed several days so President Trump's signature will be on them, the Treasury Department announces the checks are being issued "on time, as planned" and that there is no delay.

127. APRIL 25, 2020: Politico reports President Trump owes the Bank of China tens of millions of dollars on a loan coming due in 2022, as he deals with China on coronavirus. However, the Bank of China issues a statement saying it held the loan for only twenty-two days back in 2012 before selling it to a US real estate firm.

128. APRIL 28, 2020: Yahoo! News reporter Hunter Walker asks Trump a question with false information in it during an Oval Office meeting: "Overall, South Korea has done five times more tests than the U.S. per capita. Why is that?" "I don't think that's true," Trump replies. "That is true," Walker insists. In fact, South Korea's testing was 11 per 100,000 people and the United States is at 17 per 100,000. Walker later apologizes in a tweet.

129. MAY 10, 2020: NBC's Chuck Todd on *Meet the Press* uses a deceptively edited comment by Attorney General William Barr about the case of

Lieutenant General Michael Flynn. The network later apologizes for the error.

130. MAY 10, 2020: CBS's *60 Minutes* tweets that Secretary of State Mike Pompeo "attempted to resurrect a debunked theory that the virus was man-made in China." Pompeo said the opposite.

131. JUNE 2, 2020: Mediaite writes an account of demonstrations outside the White House, quoting numerous reporters as saying "tear gas" was unjustifiably used. The US Park Police and Secret Service insist no tear gas was used; they used pepper balls after a litany of violent acts some protesters had allegedly committed, including pelting officers with objects, throwing heavy objects at Attorney General Barr, and trying to grab police weapons.

Index

About the Author

Sharyl Attkisson has been a working journalist for more than thirty-five years and is host and managing editor of the nonpartisan Sunday-morning TV program *Full Measure with Sharyl Attkisson*. She has covered controversies under the administrations of Bill Clinton, George W. Bush, Barack Obama, and Donald Trump, emerging with a reputation, as the *Washington Post* put it, as a "persistent voice of news-media skepticism about the government's story." She is the recipient of five Emmy Awards and an Edward R. Murrow Award for investigative reporting. She has worked at CBS News, PBS, and CNN and is a fifth-degree black belt master in Taekwondo.